THE SUPER SOURCE®

Snap Cubes®

ETA hand2mind®

Vernon Hills, IL

ETA hand2mind® extends its warmest thanks to the many teachers and students across the country who helped ensure the success of The SUPER SOURCE® series by participating in the outlining, writing, and field testing of the materials

The SUPER SOURCE® Snap Cubes® Grades 3–4
75361
ISBN 978-1-57452-013-2

500 Greenview Court • Vernon Hills, Illinois 60061-1862 • 800.445.5985 • hand2mind.com

Printed in the United States of America.

15 16 17 18 19 20 21   15 14 13 12 11

THE SUPER SOURCE®
Table of Contents

Using THE SUPER SOURCE®

The Super Source is a series of books each of which contains a collection of activities to use with a specific math manipulative. Driving **the Super Source** is ETA hand2mind's conviction that children construct their own understandings through rich, hands-on mathematical experiences. Although the activities in each book are written for a specific grade range, they all connect to the core of mathematics learning that is important to every K–6 child. Thus, the material in many activities can easily be refocused for children at other grade levels. Because the activities are not arranged sequentially, children can work on any activity at any time.

The lessons in **the Super Source** all follow a basic structure consistent with the vision of mathematics teaching described in the *Curriculum and Evaluation Standards for School Mathematics* published by the National Council of Teachers of Mathematics.

All of the activities in this series involve Problem Solving, Communication, Reasoning, and Mathematical Connections—the first four NCTM Standards. Each activity also focuses on one or more of the following curriculum strands: Number, Geometry, Measurement, Patterns/Functions, Probability/Statistics, Logic.

HOW LESSONS ARE ORGANIZED

At the beginning of each lesson, you will find, to the right of the title, both the major curriculum strands to which the lesson relates and the particular topics that children will work with. Each lesson has three main sections. The first, GETTING READY, offers an *Overview*, which states what children will be doing, and why, and a list of "What You'll Need." Specific numbers of Snap Cubes are suggested on this list but can be adjusted as the needs of your particular situation dictate. Before an activity, cubes can be counted out and placed in containers of self-sealing plastic bags for easy distribution. If appropriate for an activity, the cubes might be distributed as color rods—10 cubes per color. When crayons are called for, it is understood that their colors are those that match the Snap Cubes and that markers may be used in place of crayons. Blackline masters that are provided for your convenience at the back of the book are referenced on this list. Paper, pencils, scissors, tape, and materials for making charts, which are necessary in certain activities, are usually not.

Although the overhead Snap Cubes are always listed in "What You'll Need" as optional, these materials are highly effective when you want to demonstrate the use of Snap Cubes. As you move the cubes on the screen, children can work with the same materials at their seats. Children can also use the overhead to present their work to other members of their group or to the class.

The second section, THE ACTIVITY, first presents a possible scenario for *Introducing* the children to the activity. The aim of this brief introduction is to help you give children the tools they will need to investigate independently. However, care has been taken to avoid undercutting the activity itself. Since these investigations are designed to enable children to increase their own mathematical power, the idea is to set the stage but not steal the show! The heart of the lesson, *On Their Own*, is found in a box at the top of the second page of each lesson. Here, rich problems stimulate many different problem-solving approaches and lead to a variety of solutions. These hands-on explorations have the potential for bringing children to new mathematical ideas and deepening their skills.

On Their Own is intended as a stand-alone activity for children to explore with a partner or in a small group. Be sure to make the needed directions clearly visible. You may want to write them on the chalkboard or on an overhead or to present them either on reusable cards or on paper. For children who may have difficulty reading the directions, you can read them aloud or make sure that each group includes at least one "reader."

The last part of this second section, *The Bigger Picture*, gives suggestions for how children can share their work and their thinking and make mathematical connections. Class charts and children's recorded work provide a springboard for discussion. Under "Thinking and Sharing," there are several prompts that you can use to promote discussion. Children will not be able to respond to these prompts with one-word answers. Instead, the prompts encourage children to describe what they notice, tell how they found their results, and give the reasoning behind their answers. Thus children learn to verify their own results, rather than relying on the teacher to determine if an answer is "right" or "wrong." Though the class discussion might immediately follow the investigation, it is important not to cut the activity short by having a class discussion too soon.

The Bigger Picture often includes a suggestion for a "Writing" (or drawing) assignment. These are meant to help children process what they have just been doing. You might want to use these ideas as a focus for daily or weekly entries in a math journal that each child keeps.

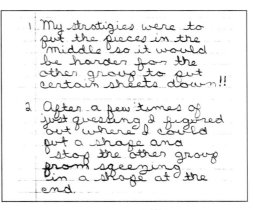

From: *Prize Inside* **From: *Tetra Fill-In***

The Bigger Picture always ends with ideas for "Extending the Activity." Extensions take the essence of the main activity and either alter or extend its parameters. These activities are particularly suited to a class that becomes deeply involved in the primary activity or to children who finish before the others. In any case, it is probably a good idea to expose the entire class to the possibility of, and the results from, such extensions.

The third and final section of the lesson is TEACHER TALK. Here, in *Where's the Mathematics?*, you can gain insight into the underlying mathematics of the activity and discover some of the strategies children are apt to use as they work. Solutions are also given—when such are necessary and/or helpful. Because *Where's the Mathematics?* provides a view of what may happen in the lesson, as well as the underlying mathematical concept that may grow out of it, you may want to read this section before presenting the activity to children.

USING THE ACTIVITIES

The Super Source has been designed to fit into a variety of classroom environments. These range from a completely manipulative-based classroom to one in which manipulatives are just beginning to play a part. You may choose to use some activities in *the Super Source* as described in each lesson (introducing an activity to the whole class, then breaking the class up into groups that all work on the same task, and so forth). You will then be able to circulate among the groups as they work to observe and perhaps comment on each child's work. This approach requires a full classroom set of materials but allows you to concentrate on the variety of ways that children respond to a given activity.

Alternatively, you may wish to make two or three related activities available to different groups of children at the same time. You may even wish to use different manipulatives to explore the same mathematical concept. (Cuisenaire® Rods and Color Tiles, for example, can be used to teach some of the same concepts as Snap Cubes.) This approach does not require full classroom sets of a particular manipulative. It also permits greater adaptation of materials to individual children's needs and/or preferences.

If children are comfortable working independently, you might want to set up a "menu"— that is, set out a number of related activities from which children can choose. Children should be encouraged to write about their experiences with these independent activities.

However you choose to use *the Super Source* activities, it would be wise to allow time for several groups or the entire class to share their experiences. The dynamics of this type of interaction, where children share not only solutions and strategies but also feelings and intuitions, is the basis of continued mathematical growth. It allows children who are beginning to form a mathematical structure to clarify it and those who have mastered just isolated concepts to begin to see how these concepts might fit together.

Again, both the individual teaching style and combined learning styles of the children should dictate the specific method of utilizing *the Super Source* lessons. At first sight, some activities may appear too difficult for some of your children, and you may find yourself tempted to actually "teach" by modeling exactly how an activity can lead to a particular learning outcome. If you do this, you rob children of the chance to try the activity in whatever way they can. As long as children have a way to begin an investigation, give them time and opportunity to see it through. Instead of making assumptions about what children will or won't do, watch and listen. The excitement and challenge of the activity—as well as the chance to work cooperatively—may bring out abilities in children that will surprise you.

If you are convinced, however, that an activity does not suit your students, adjust it, by all means. You may want to change the language, either by simplifying it or by referring to specific vocabulary that you and your children already use and are comfortable with. On the other hand, if you suspect that an activity isn't challenging enough, you may want to read through the activity extensions for a variation that you can give children instead.

RECORDING

Although the direct process of working with Snap Cubes is a valuable one, it is afterward, when children look at, compare, share, and think about their constructions, that an activity yields its greatest rewards. However, because Snap Cube designs can't always be left intact, children need an effective way to record their work. To this end, at the back of this book, recording paper is provided for reproduction. The "What You'll Need" listing at the beginning of each lesson often specifies the kind of recording paper to use. For example, it seems natural for children to record Snap Cube patterns on grid paper. Yet it is important for children to use a method of recording that they feel comfortable with. Frustration in recording their structures can leave children feeling that the actual activity was either too difficult or just not fun! Thus, there may be times when you feel that children should just share their work rather than record it.

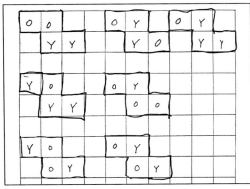

From: *Two-Color Tetras*

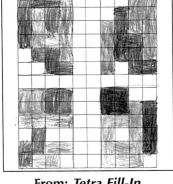

From: *Tetra Fill-In*

Young children might duplicate their work on grid paper by coloring in boxes on grids that exactly match the cubes in size. Older children may be able to use smaller grids or even construct the recording paper as they see fit.

From: *Showing One-Third*

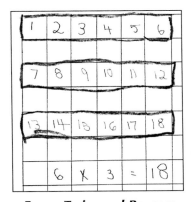

From: *Trains and Boxcars*

Another type of recording paper, isometric dot paper, can also be introduced for recording three-dimensional Snap Cube constructions. As they become more familiar with isometric dot paper, children begin to see that they can represent a corner of a cube by a dot on paper and that they now have a method of communicating visually about three-dimensional objects.

Another interesting way to "freeze" a Snap Cube design is to create it using a software program, such as *Building Perspective*, and then get a printout. Children can use a classroom or resource-room computer if it is available or, where possible, extend the activity into a home assignment by utilizing their home computers.

Recording involves more than copying structures. Writing, drawing, and making charts and tables are also ways to record. By creating a table of data gathered in the course of their investigations, children are able to draw conclusions and look for patterns. When children write or draw, either in their group or later by themselves, they are clarifying their understanding of their recent mathematical experience.

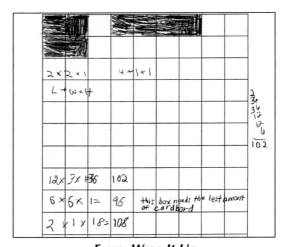

From: *The Staircase Problem*

From: *Cleared for Take-Off*

From: *Loose Caboose*

From: *Wrap It Up*

With a roomful of children busily engaged in their investigations, it is not easy for a teacher to keep track of how individual children are working. Having tangible material to gather and examine when the time is right will help you to keep in close touch with each child's learning.

Exploring Snap Cubes®

Snap Cubes are a versatile collection of 3/4-inch interlocking cubes which come in ten colors and connect on all six sides. They are pleasant to handle, easy to manipulate, and although simple in concept, can be used to develop a wide variety of mathematical ideas at many different levels of complexity. Since Snap Cubes come in ten different colors, the cubes are useful for developing patterns, both one- and two-dimensional, based on color. The cubes can be arranged in a single layer to naturally fit into a square grid pattern, or can be used to cover positions on a printed grid or game board. When the cubes are used to build three-dimensional structures, they lead naturally to the concepts of volume and surface area.

From: *Explorations with Four Cubes*

The colors of the Snap Cubes can also be used to identify cubes in other contexts. For example, the different colors can represent designated quantities in various number situations. They become a sampling device when they are drawn from a bag, and they aid in concretely building bar graphs.

From: *Prize Inside*

WORKING WITH SNAP CUBES

Snap Cubes make natural and appealing counters. Since they snap together firmly, they are useful for young children as number models. If children build a stick corresponding to each number 1 to 10, it is natural for them to arrange them in a staircase, and to talk about greater and less, longer and shorter. Numbers might also be represented by the following cube patterns, which are easily sorted into "even" ones where each cube is paired with another, and "odd" ones, where there is an "odd man out."

Snap Cubes also help children to more easily see relationships such as the following:

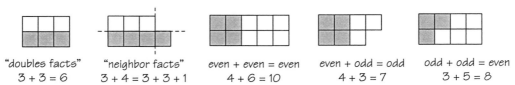

| "doubles facts"
3 + 3 = 6 | "neighbor facts"
3 + 4 = 3 + 3 + 1 | even + even = even
4 + 6 = 10 | even + odd = odd
4 + 3 = 7 | odd + odd = even
3 + 5 = 8 |

Since there are large numbers of cubes in a set of Snap Cubes, they are useful for estimation and for developing number sense. Children can make a long rod with the cubes, estimate how many there are in the rod, and then separate the rod into sticks of 10, identify how many tens they have, and count the "leftovers" to find how many ones there are.

The colors of the cubes further make them useful in developing the concept of place value. Each color can represent a place value, and children can play exchange games where if they have 10 of one color they can exchange them for one of the next color.

Snap Cubes are very suitable for developing understanding of the meaning of addition. They can be used as loose counters, with a different color for each addend. The colors can also broaden children's understanding of subtraction. Children often think initially of subtraction as "take away." To act out 6 – 4, children put out 6 cubes and take 4 away.

Snap Cubes are also ideal for developing the concept of multiplication, both as grouping and as an array. To show 3 x 4, children can make 3 "cube trains" with 4 in each, and count them all. Arranging these cubes in a rectangular array not only makes it easy to understand why 3 x 4 = 4 x 3, but also leads naturally into a model for understanding the formula for the area of a rectangle. In addition, Snap Cubes are suitable for exploring area, perimeter, volume, and surface area relations.

> 1. some are long some are short.
> 2. I counted all the cubes then I cut out the paper.
> 3. the box with 18 was the longest and had the most cubes, the 6×6×1 used the lest amont of card.

From: *Wrap It Up*

Snap Cubes are a wonderful tool to use in helping children to represent numbers in terms of factors and to understand procedures of finding greatest common divisors and least common multiples. Snap Cubes are also a natural unit for length, and can lead to early experience with ratio and proportion. Children can measure the same length in Snap Cubes and in another unit, perhaps inches. They can record their results for a few different lengths. They may then measure in just one unit, and predict the measure in the other.

ASSESSING CHILDREN'S UNDERSTANDING

Snap Cubes are wonderful tools for assessing children's mathematical thinking. Watching children work on their Snap Cubes gives you a sense of how they approach a mathematical problem. Their thinking can be "seen," in so far as that thinking is expressed through the way they construct, recognize, and continue spatial patterns. When a class breaks up into small working groups, you are able to circulate, listen, and raise questions, all the while focusing on how individuals are thinking. Here is a perfect opportunity for authentic assessment.

Having children describe their structures and share their strategies and thinking with the whole class gives you another opportunity for observational assessment. Furthermore, you may want to gather children's recorded work or invite them to choose pieces to add to their math portfolios.

By my favret noddrs ThaT are 7 10 11 12 5 3 9 8 6 becas Thay are my favnt noddrs.

12 and 1 becas you cod't get 1 and 12 was haord To rolle.

From: *Cleared for Take-Off*

Dear Tara,
This is the way you can win at nim. Try to get it down for your partener to have 3 cubs lleft. Then you can win if they take 2 or 1. Don't tell them you are trying to do it.
Your Friend,
Julie

From: *NIM Stick Strategy*

Connect THE SUPER SOURCE® to NCTM Standards.

STRANDS

	PROBLEM SOLVING	COMMUNICATION	REASONING	CONNECTIONS	Geometry	Logic	Measurement	Number	Patterns/Functions	Probability/Statistics
A TOWER OF SQUARES	◆	◆	◆	◆	◆			◆	◆	
CLEARED FOR TAKE-OFF	◆	◆	◆	◆	◆			◆		◆
EXPLORATIONS WITH FOUR CUBES	◆	◆	◆	◆	◆	◆				
GRAB BAG MATH	◆	◆	◆	◆				◆		
LOOSE CABOOSE	◆	◆	◆					◆		
MAKING FRAMES	◆	◆		◆				◆	◆	
NIM STICK STRATEGY GAME	◆	◆	◆			◆		◆		
ONE OF A KIND	◆		◆	◆	◆	◆				
ORDERING THE TETRAS	◆	◆	◆	◆	◆	◆				
PLANS AND STRUCTURES	◆	◆	◆	◆	◆					
PRIZE INSIDE	◆	◆	◆							◆
SHOWING ONE-THIRD	◆	◆	◆	◆				◆		
TAKE THE CAKE	◆	◆		◆				◆	◆	
TETRA FILL-IN	◆	◆	◆	◆	◆	◆				
THE STAIRCASE PROBLEM	◆	◆	◆	◆				◆	◆	
TRAINS AND BOXCARS	◆	◆	◆	◆				◆		
TWO-COLOR TETRAS	◆	◆	◆	◆	◆					◆
WRAP IT UP	◆	◆	◆	◆	◆		◆			

Correlate THE SUPER SOURCE® to your curriculum.

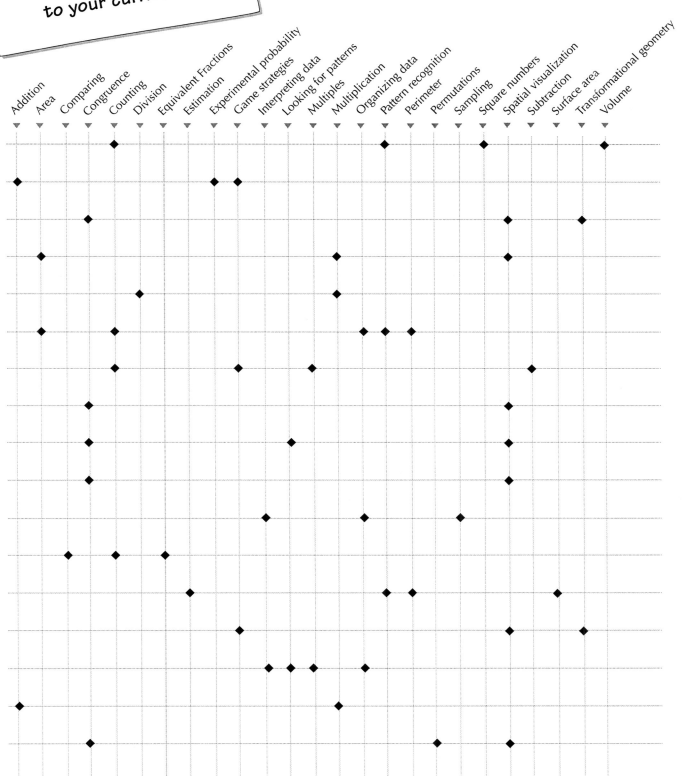

Addition · Area · Comparing · Congruence · Counting · Division · Equivalent Fractions · Estimation · Experimental probability · Game strategies · Interpreting data · Looking for patterns · Multiples · Multiplication · Organizing data · Pattern recognition · Perimeter · Permutations · Sampling · Square numbers · Spatial visualization · Subtraction · Surface area · Transformational geometry · Volume

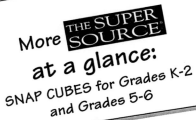

More THE SUPER SOURCE at a glance:
SNAP CUBES for Grades K-2 and Grades 5-6

Classroom-tested activities contained in these *Super Source* Snap Cubes books focus on the math strands in the charts below.

THE SUPER SOURCE® Snap Cubes®, Grades K–2

Geometry	Logic	Measurement
Number	Patterns/Functions	Probability/Statistics

THE SUPER SOURCE® Snap Cubes®, Grades 5–6

Geometry	Logic	Measurement
Number	Patterns/Functions	Probability/Statistics

More THE SUPER SOURCE at a glance:
ADDITIONAL MANIPULATIVES
for Grades 3–4

Classroom-tested activities contained in these *Super Source* books focus on the math strands as indicated in these charts.

THE SUPER SOURCE — Tangrams, Grades 3–4

Geometry	Logic	Measurement
Number	Patterns/Functions	Probability/Statistics

THE SUPER SOURCE — Cuisenaire® Rods, Grades 3–4

Geometry	Logic	Measurement
Number	Patterns/Functions	Probability/Statistics

THE SUPER SOURCE — Geoboards, Grades 3–4

Geometry	Logic	Measurement
Number	Patterns/Functions	Probability/Statistics

THE SUPER SOURCE — Color Tiles, Grades 3–4

Geometry	Logic	Measurement
Number	Patterns/Functions	Probability/Statistics

THE SUPER SOURCE — Pattern Blocks, Grades 3–4

Geometry	Logic	Measurement
Number	Patterns/Functions	Probability/Statistics

Overview of the Lessons

Snap Cubes, Grades 3–4

A TOWER OF SQUARES

- Square numbers
- Counting
- Pattern recognition
- Volume

Getting Ready

What You'll Need

Snap Cubes, about 100 per pair

Calculators, one per pair

Overview

Children use Snap Cubes to build larger and larger square prisms and stack them to form a tower. They predict the numbers of cubes needed to produce larger squares and towers. In this activity, children have the opportunity to

- ◆ investigate square numbers and growth patterns
- ◆ organize and analyze data
- ◆ use patterns to make predictions

layer 1 - 1 cube
layer 2 - 4 cubes
layer 3 - 9 cubes

The Activity

Introducing

- ◆ Display a Snap Cube and talk about its attributes, such as the number of faces, edges, and vertices.
- ◆ Snap four cubes together to form a 2 x 2 x 1 prism. Ask children to compare its attributes to the original cube shown.
- ◆ Stack the original cube on top of the 2 x 2 x 1 prism. Ask the children to predict what the next layer would look like.
- ◆ Ask a volunteer to build the next layer and add it to the tower to validate their predictions.

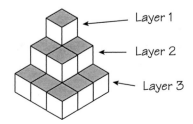

Layer 1
Layer 2
Layer 3

On Their Own

> ***How many Snap Cubes do you need to build a tower made of layers of square prisms?***
>
> - Work with a partner. Use Snap Cubes to build a tower according to this description:
>   - ◆ Layer 1 has one cube.
>   - ◆ Layer 2 is a square prism that is two cubes long, two cubes wide, and one cube high.
>   - ◆ Layer 3 is a square prism that is three cubes long, three cubes wide, and one cube high.
>
>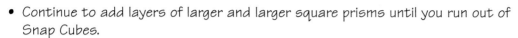
>
> - Continue to add layers of larger and larger square prisms until you run out of Snap Cubes.
> - For each layer you make, record the number of cubes in the layer and the total number of cubes in the tower.
> - Look for patterns in your recording.
> - Predict the number of Snap Cubes you would need to build a tower with 11 layers.

The Bigger Picture

Thinking and Sharing

Call children together to create a class chart that looks like this:

Layer Number	Number of Cubes in the Layer	Total Number of Cubes in the Tower
1	1	1
2	4	5
3	9	14
..	..	..
..	..	..
..	..	..

Discuss the data.

Use prompts like these to promote class discussion:

- ◆ What did you notice as you built your towers?
- ◆ What patterns did you notice in your towers?
- ◆ How did you find the number of cubes that would be needed to build the eleventh layer? The total number of cubes in the tower?
- ◆ How many cubes would be in the twelfth layer? How many cubes would be needed to build a 12-layer tower?

Writing

Ask children to write a set of directions that would help someone figure out how many cubes would be needed to build a 13-layer tower.

Teacher Talk

Where's the Mathematics?

In this activity, children use the geometric attributes of a square to investigate the concepts of square numbers. As they create the square layers of the tower, they generate the square numbers (1, 4, 9, 16, 25, 36, ...), look for patterns, and use those patterns to find the number of Snap Cubes needed for the square layers that are too large to build.

Organizing the data in a chart like this makes it easier to find and extend patterns.

Number of Layer	Number of Cubes in the Layer	Total Number of cubes in the Tower
1	1	1
2	4	5
3	9	14
4	16	30
5	25	55
6	36	91
7	49	140
8	64	204
9	81	285
10	100	385
11	121	506

Children may use different methods for finding the data. For example, some children may recognize that the number of cubes in each layer can be found by multiplying the number of the layer by itself. Other children may recognize that the differences between each successive pair of entries in the "Number of Cubes in the Layer" column are consecutive odd numbers 4 − 1 = 3; 9 − 4 = 5; 16 − 9 = 7, and so on. They may then use this pattern to find the rest of the data for the second column.

Extending the Activity

1. Have children explain whether a square layer could have 400 cubes; 500 cubes.

2. Ask children to build towers with rectangular layers instead of square layers. One such tower could have layers that are 1 x 1 x 1, 1 x 1 x 2, 1 x 1 x 3, and so on. Another such tower could have layers that are 1 x 1 x 1, 1 x 2 x 2, 1 x 2 x 3, and so on.

Children who are more visual learners may look down from the top of the tower and recognize that each layer looks like an L-shaped addition to the previous layer. The new L-shape requires two times the previous layer number plus one more cube for the corner. This L-shape is a visual way of explaining the pattern of odd number differences that occur in the data found in the "Number of Cubes in the Layer" column.

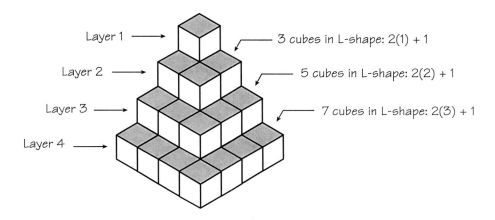

Layer 1 ⟶   3 cubes in L-shape: 2(1) + 1

Layer 2 ⟶   5 cubes in L-shape: 2(2) + 1

Layer 3 ⟶   7 cubes in L-shape: 2(3) + 1

Layer 4 ⟶

Similarly, children will have various ways of finding the data in the third column. Some children will find the entries in the "Total Number of Cubes in the Tower" column by adding the previous entry in that column to the next number in the "Number of Cubes in the Layer" column: 1 + 4 = 5; 5 + 9 = 14; 14 + 16 = 30, and so on. Others might find the entries by adding all of the entries in the "Number of Cubes in Layer" column: 1 + 4 = 5; 1 + 4 + 9 = 14; 1 + 4 + 9 + 16 = 30, and so on.

The use of calculators allows children to focus on the patterns and concepts involved in the problem, rather than on the difficulty involved in manipulating the large numbers in the last few steps of the problem.

CLEARED FOR TAKE-OFF

- Addition
- Game strategies
- Experimental probability

Getting Ready

What You'll Need

Snap Cubes, 12 per team

Dice, one pair per pair of teams

Cleared for Take-Off game board, page 91

Overhead Snap Cubes (optional)

Snap Cube grid paper transparency (optional)

Overview

Using Snap Cubes to represent airplanes on runways numbered 1 to 12, children try to clear their game board runways by rolling sums on a pair of dice that match the numbers of the runways. In this activity, children have the opportunity to

- ◆ discover that certain sums on the dice are more likely to occur
- ◆ develop strategic thinking skills

The Activity

You may want to play a full game of Cleared for Take-Off *with the children before they begin the* On Their Own.

Introducing

- ◆ Show children half of a *Cleared for Take-Off* game board showing rows 1 through 6. Explain that the rows stand for airport runways and that Snap Cubes will stand for airplanes.

- ◆ Place five Snap Cubes on the runways on one side of the game board. Call on a volunteer to place five Snap Cubes on the other side. Here is what the game board could look like:

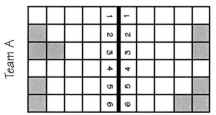

- ◆ Using a single die and the partial game board, play a demonstration game of *Cleared for Take-Off*.

On Their Own

Play *Cleared for Take-Off!*

Here are the rules.

1. This is a game for two teams. In this game, Snap Cubes stand for airplanes and the numbered rows on the game board stand for airport runways. The object of the game is to be the first team to remove all its airplanes from its side of the runway.

2. Each team takes 12 Snap Cubes and places them on the runway on its side of a game board. The placement of the airplanes is up to the teams. Some runways may have more than one airplane, and some may have no airplanes.

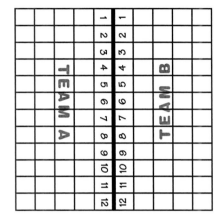

3. The teams take turns rolling the dice. Both teams check to see if they have airplanes on the runway whose number matches the sum on the dice.

4. If either or both teams have an airplane on that runway, teams may remove an airplane from their side of that runway.

5. Play continues until one team has removed all 12 of its airplanes.

- Play several games of *Cleared for Take-Off.*

- Keep a record of where you placed your airplanes and what sums came up.

- Be ready to talk about the best placement of airplanes on the runways.

The Bigger Picture

Thinking and Sharing

Call children together to discuss what happened and any strategies they discovered.

Use prompts like these to promote class discussion:

- How did you decide where to place the airplanes? Why?

- Which runways usually had airplanes left on them at the end of the game? Why?

- Do you think that putting six airplanes on runway #6 and six airplanes on runway #7 would be a good strategy? Why?

- Would you win if both teams placed their airplanes on the same runways? Why or why not?

Writing

Have children write a list of hints on how to improve a team's chances of winning *Cleared for Take-Off*.

Teacher Talk

Where's the Mathematics?

In addition to giving children practice working with sums between 2 and 12, this activity introduces children to concepts of probability. The important ideas about which sums on two dice occur most frequently develop over time. For example, the sums 5, 6, 7, 8 come up more often than 2, 3, 4, 9, 10, 11, or 12. There is no guarantee, however, that a 2 or 12 won't come up because they do have a chance of occurring. Thus, finding the right combination of Snap Cube placement is partly luck and partly based on knowing about which sums come up more often with two dice.

By asking children to look at which airplanes are left on the board at the end of the game, you will be prompting them to think about the likelihood of certain sums occurring less often in this game. Conversely, if children record which sums did occur, they might be able to use that information to place their airplanes for the next game and improve their chances of winning. When asked to explain why they think certain sums come up more frequently than others, children may point out the different combinations that can form that sum. For example, 7 = 4 + 3 or 6 + 1 or 5 + 2; but 3 can only be formed from 1 + 2. If children happened to use different-colored dice, some may point out that there are actually twice as many combinations because a red 4 and a white 3 gives a sum of 7 just as a red 3 and a white 4 will. So using different colored dice gives six ways to get a sum of 7 and two ways to get a sum of 3.

If children seem interested in exploring all the combinations, you could tally their responses on a chart like this:

RED DIE	WHITE DIE	SUM	RED DIE	WHITE DIE	SUM
1	1	2	1	5	6
1	2	3	5	1	6
2	1	3	2	4	6
1	3	4	4	2	6
3	1	4	3	3	6
2	2	4	1	6	7
1	4	5	6	1	7
4	1	5	2	5	7
2	3	5	5	2	7
3	2	5	3	4	7

Extending the Activity

Using the records they kept, have children count up how many rolls of the dice it took to clear the game board. Have them make a chart of the data. You could talk about the range, median, and mode of the data.

RED DIE	WHITE DIE	SUM	RED DIE	WHITE DIE	SUM
4	3	7	4	5	9
2	6	8	5	4	9
6	2	8	4	6	10
3	5	8	6	4	10
5	3	8	5	5	10
4	4	8	5	6	11
3	6	9	6	5	11
6	3	9	6	6	12

Children might recognize that all of the even sums have an odd number of combinations and that the odd sums have an even number of combinations.

Depending upon their experience with dice, some children may be puzzled initially that the sum of 1 never comes up even though there is a 1 on each die and on the game board. Some children may have expected higher sums than 12 so the range of 2 to 12 as sums may surprise some children.

Although the sums of 5, 6, 7, and 8 come up more frequently than other sums, placing all of one's airplanes on just those numbers is probably not the wisest strategy since any of the other sums still have a chance of coming up. Placing one's airplanes exactly the same way the other team did will end in a draw.

Some children will be able to call a winner without actually finishing the game. For example, if Team A had two planes left on runway #4 and Team B had only one plane left on runway #4, the next time a sum of 4 is rolled, Team B would win. If children can articulate the reasons for declaring a winner without actually finishing the game and both teams agree with the reasoning, encourage them to start a new game.

Continue to let the children play this game for several days after the class discussion in order to give them an opportunity to test the strategies that have been discussed and convince themselves of their validity.

EXPLORATIONS WITH FOUR CUBES

- Spatial visualization
- Congruence
- Transformational geometry

Getting Ready

What You'll Need

Snap Cubes, 32 of the same color per pair

Overview

Children explore the possible shapes that can be made using four Snap Cubes. In this activity, children have the opportunity to

- ◆ develop their spatial sense
- ◆ discover that shapes that have the same volume need not be congruent

The Activity

Use Snap Cubes of the same color so children concentrate on the shape, not the color, when deciding whether two shapes are the same or not.

If children think that these shapes are different, demonstrate by rotating or flipping that they are the same.

Introducing

- ◆ Show children a shape made of two Snap Cubes. Ask if it is possible to put two Snap Cubes together to make a shape different from the one you have shown.
- ◆ Then ask children to make as many different shapes as they can using three Snap Cubes.
- ◆ Have volunteers hold up their different arrangements and talk about them. Only two shapes are possible.

On Their Own

How many different shapes can you make with four Snap Cubes?

- Work with a partner.

- Each of you make a shape with four cubes.

- Compare your shapes.

  - If they are different, set both aside.

  - If they are the same, keep only one.

- Take four more cubes. Try to build a shape that is different from what you have already built.

- Continue building shapes until you can't make any more that are different from what you already have.

The Bigger Picture

Thinking and Sharing

When children have finished building their shapes, have a volunteer offer a shape. Ask children who think they have matching shapes to bring them up and prove that the shapes are the same. Display all the shapes in one pile. Then ask for another shape and again collect the matching shapes. Continue until all the shapes have been sorted in piles where everyone can see them.

If children are ready, you may wish to introduce vocabulary such as face, edge, *and* vertex.

Use prompts such as these to promote class discussion:

- How many different shapes are there?

- How did you compare shapes to decide whether or not they were different?

- How are these two shapes different from each other?

- (Hold up two shapes that are mirror images.) What do you notice about these two shapes?

Drawing and Writing

Have children choose their favorite shape and make a sketch of it. Ask them to give it a name and describe why they gave it that name.

Extending the Activity

1. Have the children work in pairs. The first child selects one of the shapes. While keeping the shape hidden from the view of the second child, the first child describes how to build the shape. The second child uses the description to build the shape. They then compare shapes to see if they

Where's the Mathematics?

This activity helps children focus on shape to make figures different from each other, rather than on the color of the Snap Cubes within the shape or on the way they hold the shapes.

Only eight geometric shapes are possible.

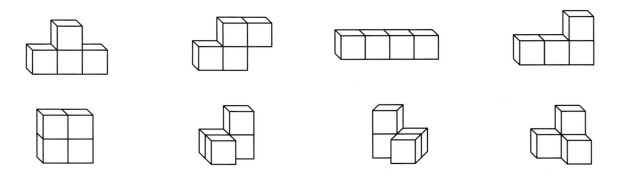

Children may or may not find all the possible geometric shapes within their individual groups or they may find "extras." For example, they may think that they have found a different shape such as a diamond simply by rotating a square of four cubes. This change in orientation might make the child think that the shapes are different, but geometrically speaking, these shapes are identical, or congruent.

are the same. Ask which shapes were easiest to describe, which were hardest to describe, and what descriptive words helped the most.

2. Have children investigate how many different shapes could be made using five Snap Cubes.

Many groups will approach the task of finding all possible shapes in a random manner. Other groups may develop a system for finding all the possible configurations. For example, they may start with four cubes in a row and then move one cube to find all the possible figures that could be made with only three cubes in the bottom row. Then they may explore all the figures possible with only two cubes in the bottom row.

Children who gain facility with comparing shapes by rotating and flipping them may feel frustrated when they compare mirror shapes such as these. Although the shapes are different, it appears that they might be the same if a certain flip or turn could be found.

Some children may need to copy the shapes other children have made to convince themselves of how the shape is configured.

This activity sets the stage for children's future work in solid geometry and provides a basis for understanding concepts—rotations, flips, slides—in transformational geometry.

GRAB BAG MATH

Getting Ready

What You'll Need

Grab Bags, 36: The first bag contains one Snap Cube; the second two; the third three; and so on to 36 cubes. *(If you don't have the 666 Snap Cubes needed to pack the bags, put a number from 1 to 36 in each bag; then children can collect and return that number of Snap Cubes from the classroom supply.)*

Large box to hold all the grab bags

Scissors

Snap Cube grid paper, page 90

Overview

Using Snap Cubes, children explore the various one-layer rectangular boxes that can be made from different numbers of cubes. In this activity, children have the opportunity to

- link multiplication with the concept of area
- organize data
- use a pattern to make predictions

The Activity

In this activity, orientation does not matter. For example, a 2 x 3 box is the same as a 3 x 2 box.

3 cubes in a row   2 cubes in a row
2 rows of cubes   3 rows of cubes

Introducing

- Mark the grab bag containing six Snap Cubes so that you can identify it, but the children cannot.
- Make a dramatic show of reaching into the box of grab bags and extracting the bag that contains six Snap Cubes. Empty the contents so children can see them.
- Ask how they could make a rectangular box, one layer high, using all six Snap Cubes.
- Ask a volunteer to describe how many cubes are in each row and how many rows are in the box. Explain that these numbers describe the dimensions of the box.
- Ask if it is possible to use six Snap Cubes to make other rectangular boxes that are one layer high.

On Their Own

Can you figure out all the ways to build 1-layer rectangular boxes with Snap Cubes?

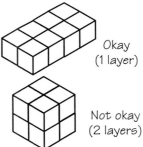

Okay
(1 layer)

Not okay
(2 layers)

- Work with a partner. Pick a grab bag from the box.

- Using the Snap Cubes in the bag, build all the rectangular boxes possible for that number of cubes. Make sure that each rectangular box has only one layer.

- Copy each rectangle onto grid paper and cut it out. Label each rectangle with its dimensions.

- Draw another grab bag and repeat the activity.

- Continue until there are no grab bags remaining.

The Bigger Picture

Thinking and Sharing

When there are no grab bags remaining, call children together. Create a class chart with columns labeled from 1 to 36. Call on volunteers to post their rectangles in the appropriate places on the chart.

Use prompts such as these to promote class discussion:

- Which rectangles have a side with two cubes? Three cubes? Four cubes? Five cubes?

- Can you predict the next four amounts of cubes that could form a rectangle with a side of two cubes? Three cubes? Four cubes? Five cubes?

- Which numbers of cubes had only one possible rectangle?

- What other patterns do you notice?

- Which rectangles are also squares?

- Do you think that if you continued this activity, you could ever find a number of cubes that couldn't be made into a rectangle? Explain.

Extending the Activity

1. Have children figure out the perimeter for each box. Ask them to describe any patterns they notice.

2. Ask children to find which numbers of cubes could be made into rectangular boxes that contain two layers.

3. Ask children to find the number of cubes between 1 and 200 that could be boxed in one-layer square packages.

Where's the Mathematics?

This activity helps children develop the connections between a number and its factors and thereby reinforces multiplication and division facts. In addition, this activity may be used as an informal introduction to the concept of finding the area of a rectangle by multiplying its dimensions.

As children explore the ways to create rectangles, some will attack the problem using trial and error, taking the specified number of cubes and randomly rearranging them until they form a rectangle. Others will have a more orderly approach, first searching for rectangles with two cubes on a side, then three, then four, and so forth. Many children will discover all the possible rectangular configurations and still try to find more because they will not know how to tell when they have exhausted all the possibilities.

This activity also develops children's spatial sense. At first, some children may not see that a 2 x 3 rectangle is the same as a 3 x 2 rectangle. They may have to physically rotate one of the rectangles to get it to match the orientation of the other rectangle before they are convinced. Some children may question whether or not a square is a rectangle.

During the class discussion, make a list of all the rectangles that have two cubes on one side (those composed of 2, 4, 6, 8, and so on up to 36 cubes). Linking this to the even numbers or multiples of two may be a revelation to some children. Likewise, those rectangles that had three cubes on a side can be linked to the multiples of 3: 3, 6, 9, 12, 15, ..., 36. All the numbers of cubes, except 1, that form only one rectangle comprise a set of prime numbers (2, 3, 5, 7, 11, 13, 17, 19, 23, 29, and 31). A prime number has exactly two factors, itself and one, and this activity very concretely demonstrates that definition. The number 1 does not belong to the set of prime numbers because it has only one unique factor.

The numbers of cubes that can be configured in square—1, 4, 9, 16, 25, 36—are appropriately called square numbers.

Writing the factors symbolically at the bottom of the class chart further
reinforces the order of the multiplication tables. Color could also be added
to the chart to emphasize patterns. For example, the multiples of 2 could be
highlighted in red and the multiples of 3 in blue. This would result in the
multiples of 6 being coded in both red and blue. The prime numbers and
the square numbers could also be assigned their individual colors.

1	2	3	4	5	6	7	8	9	10	11	12
1 x 1	1 x 2	1 x 3	1 x 4	1 x 5	1 x 6	1 x 7	1 x 8	1 x 9	1 x 10	1 x 11	1 x 12
	2 x 2				2 x 3		2 x 4		2 x 5		2 x 6
								3 x 3			3 x 4

13	14	15	16	17	18	19	20	21	22	23	24
1 x 13	1 x 14	1 x 15	1 x 16	1 x 17	1 x 18	1 x 19	1 x 20	1 x 21	1 x 22	1 x 23	1 x 24
	2 x 7		2 x 8		2 x 9		2 x 10		2 x 11		2 x 12
		3 x 5			3 x 6			3 x 7			3 x 8
			4 x 4				4 x 5				4 x 6

25	26	27	28	29	30	31	32	33	34	35	36
1 x 25	1 x 26	1 x 27	1 x 28	1 x 29	1 x 30	1 x 31	1 x 32	1 x 33	1 x 34	1 x 35	1 x 36
	2 x 13		2 x 14		2 x 15		2 x 16		2 x 17		2 x 18
		3 x 9			3 x 10			3 x 11			3 x 12
			4 x 7				4 x 8				4 x 9
5 x 5					5 x 6					5 x 7	
											6 x 6

In addition to providing practice with multiplication and division facts, this
activity lays the groundwork for children's future work in finding the
factorization of a number—a skill that will be crucial when they study algebra.

LOOSE CABOOSE

- Division
- Multiplication

Getting Ready

What You'll Need

Snap Cubes, 27 per pair
Dice, one die per pair

Overview

Children play a game in which they roll a die to determine how many trains of equal length to build from a pile of 27 Snap Cubes. They write a division sentence to describe what happens during each turn. In this activity, children have the opportunity to

- ◆ view division as making same-size sets
- ◆ practice using division symbolism
- ◆ look for patterns in division problems

The Activity

You may have to explain that the caboose is the last car on a freight train.

You may want to play a game of Loose Caboose *with children before they begin* On Their Own.

Introducing

- ◆ Show children a pile of 17 Snap Cubes. On the chalkboard, write $\overline{)17}$.

- ◆ Ask a volunteer to roll a die. Use the number that comes up to determine how many trains of equal length to build. Build that number of trains and set aside the remaining cubes, or the "loose caboose." For example, if the volunteer rolls a 3, form three trains of five cubes each and set aside the two leftover cubes.

- ◆ On the chalkboard, finish the division problem $3\overline{)17}$ R2. Identify 3 as the number of trains, 5 as the number of Snap Cubes in each train, and 2 as the remaining number of Snap Cubes, or the loose caboose.

- ◆ Now set aside the two loose caboose cubes and use the Snap Cubes that are left to write a new division: $\overline{)15}$. Have another volunteer roll the die and repeat the activity.

On Their Own

Play *Loose Caboose!*

Here are the rules.

1. This is a game for two players. The object is to wind up with more Snap Cubes.

2. Players start with a pile of 27 Snap Cubes and decide who goes first.

3. The first player writes the beginning of a division problem, $\overline{)27}$, and rolls a die to find out how many trains of equal length to build from the 27 cubes.

4. The first player builds the trains and keeps any "loose caboose" cubes that are left after the trains are built. Each of the trains should be as long as possible.

5. The first player completes the division problem. For example, if a 4 was rolled:

   $\qquad\qquad\qquad 6 \leftarrow$ Number of cubes in each train

   Number of trains $\rightarrow 4\overline{)27}\quad$ R3 $\leftarrow$ Number of loose caboose cubes
   (Number rolled)

6. If there are no loose caboose cubes, the player still completes the division.

7. The second player begins his or her turn using the cubes that are left. In the example above, there were three loose caboose cubes, so the second player would begin with 24 cubes and write $\overline{)24}$.

8. Players take turns until there are no Snap Cubes left.

- Play at least two full games of *Loose Caboose*.
- Look for patterns in the division problems.

The Bigger Picture

Thinking and Sharing

Have children post their division problems with a remainder of zero in one column. Do the same for division problems with remainders of 1, 2, 3, 4, and 5.

Use prompts like these to promote class discussion:

- What patterns did you notice?
- Which numbers could you make into two trains with no leftover cubes? Into 3 trains? 4 trains? 5 trains? 6 trains?
- What happened when you rolled a 1?
- Which numbers always had leftover cubes unless a 1 was rolled?
- Which numbers had the greatest number of ways to get a remainder of zero? What happened in the game when these numbers came up?
- Which numbers and roll of the die would give you the greatest number of loose caboose cubes in one turn?

Drawing and Writing

Ask children to write or draw a description of all the ways 24 Snap Cubes could be divided into trains of equal length.

Teacher Talk

Where's the Mathematics?

This activity provides an introduction to the concept of division as the partitioning of a set into equal-sized groups. It also conveys the meaning of the remainder. Multiplication facts are reinforced when the children start the next round of play and have to determine how many Snap Cubes are left so they can begin their division problem.

Children are likely to note that 1 is a divisor of every number; in other words, dividing by 1 always leaves a remainder of zero. Children are also likely to point out that the remainder (the number of cubes in the loose caboose) is always less than the divisor (the number of trains). Children can verify this by examining the trains and the number of caboose cubes. If the number of cubes in the loose caboose is equal to or greater than the number of trains, then each of the trains can be made longer.

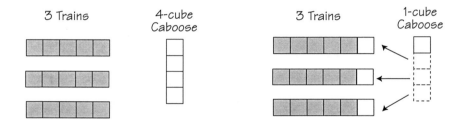

Children will notice that the numbers 2, 4, 6, 8, 10, 12, 14, 16, 18, 20, 22, 24, and 26 can be made into two trains of equal length with no remainders. They will recognize these numbers as even numbers and perhaps supply their own definition of even numbers as "the doubles" or "numbers that make equal trains with no leftovers when you roll a 2."

When they look at the numbers that have 3 as a divisor and no remainder —namely, 3, 6, 9, 12, 15, 18, 21, 24, and 27—children will be reminded of the multiplication table for 3. They may notice that this list includes every third number from the list of even numbers and has an odd-even pattern.

Extending the Activity

Have children play the game with one of these variations:
• Use an 8-sided or 10-sided die.
• Use a standard die labeled with larger numbers.
• Start with a different number of Snap Cubes.

Four is a divisor for 4, 8, 12, 16, 20, and 24. These numbers are made up of every other number from the even list. As the size of the divisor increases, the list of numbers shortens. By the time children get to the numbers that have 5 as a divisor, there are only five: 5, 10, 15, 20, and 25. The list for 6 is even shorter with only four numbers: 6, 12, 18, and 24.

Children are apt to report that 12 and 24 are the numbers they got "stuck on" in the game. By this, they mean that any roll of the die, except 5, resulted in trains with an equal number of cars and no leftover cabooses; so play went back and forth between the two players with no one winning any loose caboose cubes until a 5 was rolled.

When the children look for numbers that always give loose caboose cubes (unless a 1 was rolled), they will find 7, 11, 13, 17, 19, and 23. Later, children will learn that these numbers are part of the set called prime numbers, which have exactly two divisors: the number itself and 1. The prime numbers 2, 3, and 5 would not belong on the children's list because these numbers have no leftovers when the numbers themselves (2, 3, or 5) are rolled.

By a lucky roll of the die, or if the children have played the game enough to be able to compile exhaustive lists of data, they will see that the numbers 11, 17, and 23 hold the potential for rolling a 6 on the die and netting the largest number (5) of loose caboose cubes possible for this game.

In analyzing the parts of their division problems—namely, the dividend, divisor, quotient, and remainder—children get their first taste of the study of number theory and a foundation for dealing with division in an algebraic context.

MAKING FRAMES

- Counting
- Area
- Perimeter
- Patterns
- Organizing data

Getting Ready

What You'll Need

Snap Cubes, 10 of one color and 30 of another color per pair

Snap Cube grid paper, page 90

Overhead Snap Cubes (optional)

Overview

Children use Snap Cubes to make frames that go around various rectangular prisms. In this activity, children have an opportunity to

- ◆ find perimeter in a variety of ways
- ◆ become familiar with various rectangular shapes
- ◆ explore the relationship between area and perimeter

The Activity

Introducing

- ◆ Show children a square shape made from four Snap Cubes of the same color. Remind children that this is a 2-by-2 square and that its dimensions are written as 2 x 2.
- ◆ Ask children to imagine that the square represents a picture and you want to build a frame around the picture. Ask children to guess how many Snap Cubes it would take to make the frame.
- ◆ Then use a different color of Snap Cubes and build the frame around the square. Ask children to count with you to determine how many Snap Cubes it took to make the frame.

Number of cubes
in frame = 12

On Their Own

How can you predict the number of Snap Cubes it will take to make a frame around a Snap Cube picture?

- Work with a partner.

- Use from 1 to 10 Snap Cubes of one color to form a flat rectangular picture. Then make a frame around the picture using Snap Cubes of a different color.

- Draw your picture and its frame on grid paper.

- Record the dimensions of the picture, the number of Snap Cubes in the picture, and the number of Snap Cubes in the picture's frame. For example, suppose you made this picture with six cubes.

Dimensions of picture = 2 x 3
# of cubes in picture = 6
# of cubes in frame = 14

- Remember that you may be able to make more than one picture with the same number of cubes. For example, with six cubes you can make a picture with dimensions of 2 x 3 or a picture with dimensions of 1 x 6.

- Make several different pictures and frames, using different numbers of cubes. Record the results for each.

- Look for patterns in your recordings that will help you to predict the number of cubes in the frame if you know the dimensions of the picture.

The Bigger Picture

Thinking and Sharing

Call children together and ask them for ideas on the best way to organize their data in order to see patterns emerge. They may suggest a chart where the numbers of cubes needed to make the picture are arranged in ascending order, or they may suggest an arrangement where all the first dimensions are the same and the second dimensions increase by one. Children might also suggest pulling all the square pictures into a separate list and leaving the non-square pictures grouped together. Once children have chosen a chart format, invite them to fill in the chart with their data.

Use prompts like these to promote class discussion:

- How does this method of organization help you see patterns?

- What patterns do you see?

- Is there a way to predict the number of cubes needed for a frame if you know the dimensions of the picture? What is it?

- What did you notice about pictures that are made out of four, six, and nine cubes?

- Why do all the frames require an even number of cubes?

- If a picture measured 10 x 13, how many cubes would you need for the frame?

Drawing and Writing

Ask children to choose a number of Snap Cubes and write a set of directions for forming pictures from those Snap Cubes and an explanation of how to form frames for those pictures. Have them illustrate their directions with drawings.

Where's the Mathematics?

Children will visualize the number of cubes needed for a frame in a variety of ways. Some will explain that they add the number of cubes used across the top of the picture to the number of cubes used across the bottom of the picture, two cubes to each side dimension to account for the corners, and add all those numbers up. Applying this reasoning to a 10 x 13 picture would mean that they would need 10 for the top, 10 for the bottom and 15 on each side. Therefore, the number of cubes in the frame is 10 + 10 + 15 + 15 or 50. Others may use the conventional formula for perimeter (P = 2L + 2W) without even knowing the formula and then add 4 more cubes for the corners. Children who see it this way would explain that a 10 x 13 picture would require (2 x 10) + (2 x 13) + 4 or 20 + 26 + 4 or 50 cubes.

All of the perimeters require an even number of cubes because the same number of cubes are needed on the top of the frame as on the bottom, and likewise, the same number are required for the right side of the frame as for the left. This doubling effect leads to the even numbers. Some children may note that compact pictures require fewer cubes to frame them than long pictures. For example, a picture made of four cubes requires 12 cubes to frame it if arranged in a 2 x 2 square but 14 cubes to frame it if arranged in a 1 x 4 rectangle. The difference is even more dramatic when the numbers get larger. A 5 x 5 picture requires 24 cubes to frame it while a 1 x 25 picture requires 56 cubes! In a similar vein, children may observe that longer, skinnier rectangles, such as a 1 x 8, require more framing cubes than shorter, more compact rectangles such as a 2 x 4.

Answers to the question about the numbers 4, 6, and 9 might include, "There are two possible frames for four, six, and nine cubes," or "We can have two different shapes," and so on. Cubes in sets of four, six or nine may be arranged as rectangular arrays in several ways because they have factors other than one and themselves. In other words, 4 may be arranged as 2 x 2 as well as 1 x 4; 6 as 2 x 3 as well as 1 x 6; and 9 as 3 x 3 as well as 1 x 9. Because there are more ways to arrange the cubes, there are more ways to build the frames for them.

Extending the Activity

Have children choose a picture dimension that is not on the class chart. Ask them to predict the number of cubes in the frame based on the patterns in the chart. Then have them draw the picture and its frame to verify.

Encouraging children to come up with ways to organize the data into a chart may lead to multiple charts of the data and the opportunity to discuss the pros and cons of different methods.

Regardless of which kind of chart they elect to use, children should readily notice that each succeeding frame size is two more than the preceding (except for numbers that can be organized in more than one way). Children can extend the chart to predict what will happen with the larger numbers.

DIMENSIONS OF PICTURE	NUMBER OF CUBES IN PICTURE	NUMBER OF CUBES IN FRAME
1 x 1	1	8
1 x 2	2	10
1 x 3	3	12
2 x 2	4	12
1 x 4	4	14
1 x 5	5	16
1 x 6	6	18
2 x 3	6	14
1 x 7	7	20
1 x 8	8	22
2 x 4	8	16
1 x 9	9	24
3 x 3	9	16
1 x 10	10	26
2 x 5	10	18

This activity touches upon some number theory as well as work with area, perimeter, simple addition, and multiplication.

NIM STICK STRATEGY GAME

- Counting
- Subtraction
- Multiples
- Game strategies
- Mental arithmetic

Getting Ready

What You'll Need

Snap Cubes, 10 per team

Overview

Children play a game in which they take turns removing one or two Snap Cubes from a cube stick. They look for strategies that will enable them to take the last cube. In this activity, children have the opportunity to

- develop strategic thinking skills
- use subtraction
- recognize multiples of three
- do mental computation

Game of NIM Stick Strategy
Rules:
1. Players of teams take turns removing one or two cubes from the stick.
2. The winner is the player or team that takes the last turn.

The Activity

Introducing

- Show children a stick made of seven Snap Cubes.
- Call on a volunteer to demonstrate a game of *NIM* with you. Explain that players take turns removing one or two cubes from the stick until there are no cubes left. The winner is the person who removes the last cube(s).
- Play several more games with other volunteers.
- Go over the game rules given in *On Their Own*.

On Their Own

The Bigger Picture

Thinking and Sharing

After children have had an opportunity to play the game many times, invite them to talk about their games and describe some of the thinking they did.

Use prompts such as these to promote class discussion:

- Is there a best first move? If so, what is it? Why is it a best first move?

- Does it matter who goes first? Why do you think that?

- Do you have other good moves? What are they? Why are they good moves?

- Did you make any moves you wanted to take back? Why?

- How did you decide whether to take one or two cubes?

- Did you have a moment of surprise during a game? Try to describe that moment.

- Did anyone develop a strategy that will always work? Tell about it.

- What kinds of things did you think about when you planned your moves?

Writing

Have children write to a friend and explain the best way to win at the *NIM Stick Strategy* game. Ask them to use drawings to illustrate their explanations.

Extending the Activity

1. Have children play the game using more than ten Snap Cubes. Ask them to describe winning strategies.

2. Have children repeat the game with this change: the team that removes the last cubes loses. Ask children to compare strategies for winning this game with the strategies for winning the original game.

Teacher Talk

Where's the Mathematics?

Although this game may be played by individuals, having the children play in teams of two encourages children to communicate mathematically as they discuss and develop their winning strategies.

This game helps children develop, analyze, and compare strategies designed to produce a given outcome in a game situation. Most children will begin without any strategy in mind. As they play, they may test a variety of strategies. Some children will at first try a relatively random strategy, explaining, for example, that to win, "You should play any number until there are fewer than five cubes left. Then play very carefully."

Others may test strategies that sound systematic, but are not based on an analysis of the situation. Copying what the other team does or thinking that the first team to play is always the winner (or the loser) are examples of this type of strategy. Some children will develop a strategy, test it once successfully, and be convinced that it will always work. The class discussion will help them to see that one success is not always enough to judge the validity of a strategy.

Some children may describe a winning strategy based on the concept of multiples. The team that has the advantage is the one that removes cubes so that a multiple of three (3, 6, or 9) is left on the stick.

Here is an example of how the team that goes first can control the game and win by keeping track of sums of three:

3. Have children repeat the game with this change: they may remove one, two, or three cubes at a time. After they have played several times, ask them to compare strategies with those for the original game.

The team that goes first takes one cube.
That leaves nine cubes for the team that goes second.
Then:

If the team that goes second takes	The team that went first takes	Sum of moves	Cubes left on stick
1	2	1 + 2 = 3	6
2	1	2 + 1 = 3	3
1	2	1 + 2 = 3	0

As they describe their strategies, ask children to explain how they arrived at them, rather than just tell whether or not a particular strategy works. What is most important in this activity is developing the ability to reason out a situation and think ahead to predict the consequences of a particular action.

Some children will have difficulty articulating what they are thinking. Other children will be able to explain their winning strategy but be unable to accept that another team has expressed a similar idea in different words. Over time, this will change.

Playing the game again after changing the rules, as suggested in *Extending the Activity*, is a good way to help children learn to generalize as they adapt their winning strategy to a new situation.

ONE OF A KIND

- Spatial visualization
- Congruence
- Following directions

Getting Ready

What You'll Need

Snap Cubes, about 30 per pair
Lunch bags (optional)

Overview

Children create a coded plan and a Snap Cube structure to match that plan. Then they create a description which, when combined with the coded plan, will produce an identical structure. In this activity, children have the opportunity to

- ♦ develop writing skills in a math setting
- ♦ use precise mathematical language
- ♦ use information to build a unique three-dimensional structure

The Activity

Before children work on this activity, you may want them to do Plans and Structures *(page 54) in which children create and read coded plans for Snap Cube structures.*

Introducing

- ♦ Display this coded plan. | 3 | 1 |
- ♦ Explain that it represents a bird's eye view of a Snap Cube structure that has a tower of three cubes next to a single cube.
- ♦ Ask children to build a Snap Cube structure that would match this coded plan.
- ♦ There are three possible different structures for the coded plan. If the class does not produce all three models, be prepared to show them.

- ♦ Pick one of the three structures and hold it up to show the class.
- ♦ Ask children for suggestions on how to give information—in addition to the coded plan—so that everyone would know that this one was the structure you wanted.

On Their Own

Can you find a way to describe a Snap Cube structure so that someone else could build a structure that is an exact match of yours?

- Work with a partner. Create a coded plan that requires eight to 15 Snap Cubes. A coded plan tells how many Snap Cubes are in each tower of a structure.

- Build a structure that matches your coded plan. Here is an example of a coded plan and a matching structure.

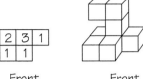

Front   Front

- Use words, symbols, and/or pictures which, when combined with your coded plan, give a description of your structure.

- Test your coded plan and description. Make sure that they describe the structure you built and no other.

- Put the coded plan and description together and hide your structure.

- Exchange coded plans and descriptions with another pair of partners. Use their coded plan and description to build their structure.

- When you are ready, compare the structure you built with the original structure the other pair built. Are they exactly the same? If not, discuss why.

The Bigger Picture

Thinking and Sharing

After pairs of children have had a chance to compare their work, invite volunteers to share their descriptions and structures.

Use prompts like these to promote class discussion:

- What kind of information needs to be included in the description?

- What was difficult about creating a description? What was easy?

- What was difficult about using someone else's description? What was easy?

- When you looked at the other pair's coded plan and description, what information was the most helpful? most confusing?

- If you did the activity again, what would you do differently?

- How did your description differ from someone else's?

Writing

Ask children to write a description of a structure with this coded plan.

1	4	2

Teacher Talk

Where's the Mathematics?

This activity of creating precise descriptions can help children strengthen their spatial reasoning skills as they search for meaningful ways to convey information about a three-dimensional object through the two-dimensional coded plan and a verbal description.

Children may come up with a variety of ways to give the required information. Some children, for example, may describe each layer of the figure with a coded plan.

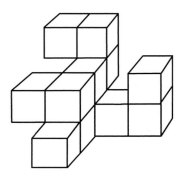

Snap Cube structure

1	2	O	O
O	2	1	2
1	2	O	O
O	1	O	O

Coded plan

O	O	O	O
O	1	1	1
O	1	O	O
O	1	O	O

Bottom layer

O	1	O	O
O	1	O	1
1	1	O	O
O	O	O	O

Middle layer

1	1	O	O
O	O	O	O
O	O	O	O
O	O	O	O

Top layer

13	14	15	16
9	10	11	12
5	6	7	8
1	2	3	4

Some children may set up a 4 x 4 grid and assign each square a number. They may then describe the structure above like this: The bottom layer has a cube on 2, 6, 10, 11, and 12; the second layer has cubes over 5, 6, 10, 12, and 14; the third layer has cubes over 13 and 14.

Extending the Activity

Ask children to build as many structures as they can that have this coded plan.

0	2	4
3	2	1

Yet another description might be as follows: The bottom layer is an L-shape with five cubes—two cubes going in each direction from the corner cube; the middle layer has four cubes connected in an L-shape that has three cubes in a straight line and one cube jutting out; the middle layer also has one cube by itself; the top layer has only two cubes that are snapped together.

Others might draw pictures, use symbols such as arrows, or abbreviations such as "F" to mean "in front of." Other children may use color to help with their descriptions. For example, "Put a green cube on top of a red cube."

Some might create coded plans for the front and side views in addition to the coded plan they already have for the top view. Using the Snap Cube structure from the previous page:

1	2	0	0
0	2	1	2
1	2	0	0
0	1	0	0

Coded plan
top view

1	1	0	0
1	3	0	1
0	3	1	1

Coded plan
front view

0	0	0	2
0	2	2	1
1	1	3	0

Coded plan
right side view

Whichever method of description children choose, they will be developing many skills with both everyday and mathematical applications. Their initial use of the coded plan requires skills that are related to the reading of maps and floor plans. As they attempt to formulate directions that can result in only one structure, children gain appreciation for and experience with using precise language.

ORDERING THE TETRAS

- **Spatial visualization**
- **Sorting**
- **Congruence**
- **Looking for patterns**

Getting Ready

What You'll Need

Snap Cubes, four of the same color, per pair

Tetra cards, one set per pair, page 92

Overview

Using Snap Cubes to verify their work, children order tetra cards so that each pattern displayed can be changed to the next one by moving just one cube. In this activity, children have the opportunity to

- ◆ build three-dimensional figures from two-dimensional drawings
- ◆ look for patterns in data
- ◆ check for congruence

The Activity

Introducing

- ◆ Display these two Snap Cube structures. Ask children to build the first one.

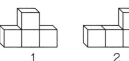

- ◆ Ask a volunteer to show how to build the second structure by moving just one cube in the first structure. Have the class build the second structure in this way. Ask if there is another way to build the second structure by moving just one cube in the first structure.

- ◆ Next, show these two structures. Establish that it is impossible to build the second structure by moving just one cube in the first.

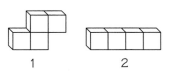

- ◆ Tell children that structures made with four Snap Cubes are sometimes called *tetras*. Explain that "tetra" comes from a Greek word meaning "four."

- ◆ Display a set of tetra cards and explain that these show all the possible tetras.

On Their Own

Can you arrange a set of tetra cards so that each structure shown can be changed to the next one by moving just one Snap Cube?

- Work with a partner. Use a set of Tetra cards like the ones shown.

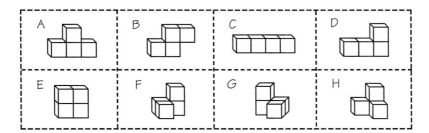

- Choose a way to arrange all eight tetra cards in order from left to right.

- Test your arrangement by building the tetra shown on the first card on the left. Then try to build the tetra shown on the second card by moving just one cube in the first tetra. Next, build the tetra shown on third card by moving just one cube in the second tetra. Continue until you have made all the tetras pictured from left to right in your arrangement.

- If you are unable to make the next tetra by moving just one cube, think of a way to rearrange the tetra cards.

- When you find an arrangement of cards that is a solution, record it.

- Look for other arrangements that are solutions and record them.

The Bigger Picture

Thinking and Sharing

Once each pair has found at least three arrangements, call children together to create a class chart. First call on volunteers to share any arrangements they found that started with tetra A. Place these in a column. Then create columns for arrangements starting with tetra B, tetra C, and so on through tetra H. Have volunteers write solutions in these columns. Allow children time to examine the chart for patterns.

Use prompts such as these to promote class discussion:

- Are there any groups of cards that appear together?

- Into which tetras can you change tetra A? tetra B? tetra C? D? E? F? G? H?

- Are there any tetra cards that always fit between two other cards?

- Does any tetra card start or end the line more often than other cards? Which ones?

- Is there any tetra card that never begins or ends the line? Do you think this is accidental or does this happen for a reason?

Writing

Have children write a note to a third or fourth grade teacher in which they describe some hints or tips to give students to help them solve this problem.

Where's the Mathematics?

This activity helps children focus on how small changes can make two structures different or non-congruent. They gain experience with making flips and rotations to move to the next structure.

Here are the tetras and some of the many solutions to the problem.

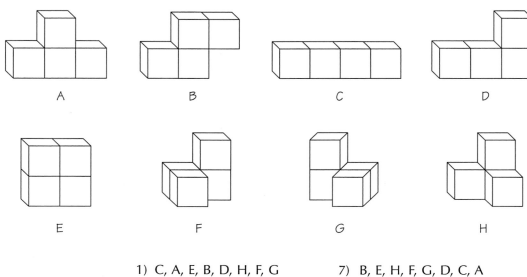

1) C, A, E, B, D, H, F, G

2) C, D, B, A, E, F, H, G

3) C, D, B, A, E, F, G, H

4) E, D, C, A, F, G, H, B

5) F, G, H, B, A, C, D, E

6) A, B, E, H, F, G, D, C

7) B, E, H, F, G, D, C, A

8) D, E, B, H, G, F, A, C

9) G, E, B, H, D, F, A, C

10) G, E, B, H, F, A, C, D

11) C, A, G, E, B, H, D, F

Extending the Activity

Have the children arrange the eight cards in a circle so that it makes a continuous loop and each card may be changed to the one on either side of it if one cube is moved.

Tetras A and D can be changed into all of the other tetras by moving just one cube. Tetra B can be changed into A, D, E, F, G, and H. Tetra B cannot be changed into C in fewer than two moves. In Tetra C, the linear arrangement and lack of right-angle connections make it difficult to fit into the line-up. C must come at the beginning or end of an arrangement or be sandwiched in between A and D (A-C-D or D-C-A).

Some children will recognize the commutative nature of the line-up. Since C cannot be changed into E, E cannot be changed into C. Tetras E, F, G, and H may be changed into any other tetra except C.

Any tetra card may begin or end the line. Children will see that new solutions may be found by moving a cluster of cards and fitting the rest of the cards in around the cluster. An example of this is the cluster E-B-H in solutions #10 and #11 in the list on the previous page. The cluster F-G-H is found in many of the arrangements in the partial list of solutions. By slightly changing the cluster to F-H-G, a whole new solution results as shown in solutions #2 and #3.

Some children will work haphazardly without any plan in mind. Other children may try to find all the solutions that begin with whatever tetra card they have decided upon. Some children will find an arrangement and then write it backwards as a second solution. For example, B-E-H-F-G-D-C-A becomes A-C-D-G-F-H-E-B. Some children will find an arrangement, such as B-E-H-F-G-D-C-A and then try to make minor changes in the arrangements such as B-E-F-H-G-D-C-A where they have interchanged the F and H but left the rest of the line alone. Sharing these different approaches helps children become more flexible in their thinking.

Working with problems where the order matters is a good foundation for probability problems involving permutations. Also, discovering that a problem may have multiple solutions encourages children to approach other problems looking for multiple solutions.

PLANS AND STRUCTURES

Getting Ready

What You'll Need

Snap Cubes, about 15 per child

Trays or large box tops from copy paper boxes

Snap Cube grid paper, page 90

Overview

Using Snap Cubes, children make structures and then create coded plans for their structures. Children exchange their structures and coded plans with other groups of children and match each structure with its coded plan. In this activity, children have the opportunity to

◆ represent a three-dimensional figure in two dimensions

◆ use a two-dimensional drawing to build a three-dimensional structure

◆ recognize the value and limitations of a coded plan

The Activity

Introducing

◆ Show children the structure to the right and have them build it.

◆ Ask children to stand up and look down at the structure. Have volunteers describe the "bird's eye view" of the structure.

◆ Draw a rectangle as shown and have children confirm that it is a bird's eye view of the structure.

◆ Put a number in each box of your rectangle.

◆ Establish that each number tells how many cubes have been used in that column of the structure.

◆ Tell the class that the rectangle you've drawn is called a *coded plan*.

◆ Display this structure and its coded plan.

◆ Discuss how this coded plan represents the structure.

On Their Own

Can you match a coded plan to its Snap Cube structure?

- Work in a group. Each of you use eight to 15 Snap Cubes to build a structure.

- Draw a coded plan of your structure and put your name on it.

- Here is an example of a 6-cube structure and its coded plan. The numbers in the coded plan tell how many cubes are in each column of the structure.

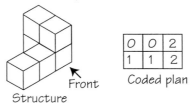

0	0	2
1	1	2

Coded plan

Structure   Front

- Have your group check that your coded plan matches your structure.

- When all the structures and coded plans have been checked, put all the structures on a tray. Put all of the coded plans in a pile and place that on the tray, too.

- Switch trays with another group.

- Try to match the structures and the coded plans. If you don't agree, talk to the creator of the coded plan.

- Then switch trays with other groups. Each time, try to match their structures and their coded plans.

The Bigger Picture

Thinking and Sharing

Once children have had the chance to match the structures and coded plans of at least two other groups, call them together to talk about what happened.

Use prompts such as these to promote class discussion:

- What strategies did your group devise to help you make the matches?

- What information on the coded plan helped you decide which structure matched that plan?

- Could you match more than one structure to the same coded plan? Explain.

- What information is missing from a coded plan?

- If a structure was turned on its side, would the coded plan change? Why?

Writing

Ask children to describe what the structure built from this coded plan could look like.

1	4

Teacher Talk

Where's the Mathematics?

Creating structures with Snap Cubes helps children develop their spatial skills. This activity gives children the opportunity to use mathematics as a means to organize and convey information in an efficient manner. It also gives children a tool for bridging the three-dimensional world around them with the two-dimensional world of paper.

Children use a variety of strategies to match the coded plans to the structures. Some add up the numbers in the coded plans to find the total number of Snap Cubes used in the structure and then search for structures with that number of Snap Cubes. Others look at the dimensions of the coded plan rectangle and then look for structures with similar dimensions. Some groups look over all of the coded plans and structures before they begin to make any matches. Other groups immediately assign each member a structure or coded plan to look for.

During their investigations, children may discover that more than one structure can be created using the same coded plan. The coded plans used in this lesson tell how many cubes are in each column, but the plans do not specify the placement of the cubes within the column. For example, the coded plan | 3 | 1 | could be used to build any of these three structures:

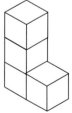 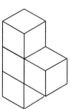 

Discussing the limitations of coded plans will set the stage for the lesson One of a Kind *(page 46) where children are asked to create a better way of describing three-dimensional figures.*

Extending the Activity

1. Have children create a coded plan and build as many different structures as they can that match the coded plan.

2. Ask children to write a description of their structure that does not use a coded plan. Make a classroom list of all the words and phrases they use to indicate location, such as *on top of, behind, to the right.*

This discovery enables children to see the need for giving additional information in order to produce a specific structure from a coded plan.

This activity can help children strengthen their spatial reasoning skills. For example, many children will not immediately recognize that the coded plan

| 3 | 1 | could be matched with a structure that looks like this:

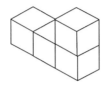

Only when the structure is turned, as shown below, does it become apparent that there is a match.

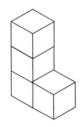

As children make the matches between coded plans and structures, they will have to re-orient the structures and communicate to their peers why the match between plan and structure now works.

PRIZE INSIDE

Getting Ready

What You'll Need

Snap Cubes, five per pair, each cube a different color

Paper bags, one per pair

Overview

Children use Snap Cubes to conduct a simulation to find the most likely number of cereal boxes they need to buy in order to collect a complete set of prizes hidden inside. In this activity, children have the opportunity to

- ◆ use simple objects to model a real-world situation
- ◆ organize and analyze data
- ◆ make predictions based on data collected from a simulation

The Activity

If children have trouble understanding how the model represents the real-life situation, discuss how the stickers and cubes are similar (three of each, each one different) and how buying a box of cereal without knowing which sticker is inside is like reaching into a paper bag without looking.

Introducing

- ◆ Describe the following scenario to children: A cereal company decides to include a free animal sticker inside each box of cereal. There are three different kinds of stickers.
- ◆ Ask children how many boxes of cereal they think they will have to buy to get a complete set of three different stickers.
- ◆ Explain that mathematicians often simulate, or model, situations like this using simple objects like cubes or counters.
- ◆ Display a paper bag and three Snap Cubes, each a different color. Tell children that you will use these to perform a simulation of the cereal box problem. Put the cubes in the bag. Ask a volunteer to reach into the bag without looking and take out a cube. Show the cube picked, record the color, and then replace the cube in the bag and shake the bag. Repeat until all three colors have been chosen.
- ◆ Ask volunteers what they think would happen if they did the experiment again and to explain their thinking.

On Their Own

A cereal company decides to put a trading card in each box of cereal it sells. There are five different trading cards. The company sends an equal number of boxes with each card to every supermarket that sells its cereal. How many boxes of cereal do you think you will need to buy in order to have a good chance of getting all 5 trading cards?

- Work with a partner.

- Put 5 different colored cubes in the bag. Pick a cube from the bag and record the color. Return the cube to the bag and shake the bag.

- Continue picking cubes and returning them to the bag until you have picked all 5 colors. Record the number of picks you made.

- Run the simulation as many times as possible.

- Decide how many boxes of cereal you think you will need to buy in order to get all 5 trading cards. Be ready to explain why you chose that number.

The Bigger Picture

Thinking and Sharing

When each pair has had a chance to run the simulation several times, ask children for suggestions on the best way to share the findings of the class. Some may suggest an oral report; others might suggest putting all the numbers they recorded on the chalkboard; still others might suggest a graph. Discuss the pros and cons of each suggestion and then make a decision. Allow time for children to organize and discuss all the data and to adjust their conclusions if they want.

Use prompts like these to promote class discussion:

- Were you more likely to choose one color cube than another? Why or why not?

- What was the least number of picks it took to get all five colors? Do you think a lesser number would be possible?

- What was the greatest number of picks it took to get all five colors? Do you think a greater number would be possible?

- Did you find it helpful to organize the data before deciding how many boxes you think you would need to buy? If so, how did you organize it?

- When you saw the results from the other groups, did you adjust the number of boxes you had decided on? If so, how did you adjust it?

- Why might it be helpful to perform the simulation more times?

Writing

Tell children that the cereal company president wants to have 100 different trading cards. Have them write to the company president encouraging or discouraging her from doing this and giving their reasons why.

Extending the Activity

1. Have children describe what they would expect the simulation results to be if there were eight trading cards instead of five. Would they expect

Where's the Mathematics?

The process of relating a problem to an experiment introduces children to mathematics as a powerful tool that they can use to analyze certain real-life situations.

This activity gives children a chance to learn the value of a simulation. A simulation produces data that reflects the actual outcomes of a real-life situation. In theory, purchasing 13 to 14 boxes should give a complete set of cards.

Since the Snap Cubes are picked from the bag at random without the child looking, each cube has an equally likely chance of being picked each time. The fact that a cube has already been picked once or twice before does not make it less likely to be picked than a cube that has not yet been picked. Given this, the range for the number of picks it takes to choose all five cubes can vary considerably. Although the fewest number of picks possible is five, chances are low that any pair of children will get this result. It is also possible, but unlikely, that a pair of children will not pick all five cubes in the amount of time allowed for this activity.

Sharing the different ways in which the children recorded their data can help them see the power of mathematics as an organizational tool. Some children will write down the name of each color as the cube is chosen. Others may write down the five colors and then use tally marks next to each color to record what is happening. Still others may record the color's name the first time it shows up and then just use tally marks for other picks that do not introduce any new colors. When the whole class shares its data, the idea of a frequency distribution graph may be suggested even though the children don't have the formal vocabulary for it. If your class has done many simulations, the frequency distribution graph will take on a bell curve shape.

the number of picks to be greater or less than the number of picks with five cards? If they were to do the simulation, could the total number of picks be less than what they got for five cards?

2. Have children describe how they would set up a simulation in which there were two trading cards possible, but there were twice as many of one card as the other.

3. Ask children how much they would expect to pay in order to get a complete set of trading cards if each box of cereal costs $3.49.

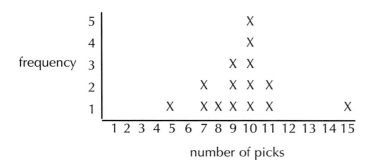

Depending on how they choose to interpret "good chance," children may elect different ways to answer the question of how many boxes of cereal they think they would need to buy to get all five trading cards. Some may answer with a range of numbers, indicating the least and greatest number of picks it took to accomplish the task. Other children may describe an "average" number of picks, possibly by selecting the middle value of a range of data (median). Some children may use the number of picks that appeared most often (mode). Be ready to accept all answers for which the children can give a reasonable justification.

When children look at the simulation results from other groups as *one* collection of data, they can discover another important property of simulations—increasing the number of trials increases the information they have on which to base their predictions, and it may show them outcomes they may not have thought were possible. It may also bring their solutions closer to the theoretical results.

SHOWING ONE-THIRD

- Counting
- Comparing
- Equivalent fractions

Getting Ready

What You'll Need

Snap Cubes, 20 each of two different colors per pair or small group

Snap Cube grid paper, page 90

Crayons

Overhead Snap Cubes and/or Snap Cube grid paper transparency (optional)

Overview

Using Snap Cubes of two different colors, children make rectangles with one-third of the one color and two-thirds of the other color. In this activity, children have the opportunity to

- ◆ discover that there are many ways to show one-third
- ◆ develop strategies for showing one-third using geometry and number
- ◆ work with ratio and proportion

The Activity

You may wish to write the fraction

number of blue cubes

total number of cubes

to help children see that the denominator of the fraction is not just the number of yellow cubes but rather all the cubes in the rectangle.

Here are some examples of what children may produce:

 2/6   3/9

 2/6

Introducing

- ◆ Show children a rectangle made of one blue cube and two yellow cubes. Ask them what fraction of the rectangle is blue.
- ◆ Display these three rectangles to establish that the location of the colors does not matter. In all three of these rectangles, one out of every three cubes is blue.

- ◆ Ask some volunteers to copy these shapes onto grid paper, color in one square to show the one-third, and cut them out. Cluster these shapes together on butcher paper and label them 1/3.
- ◆ Ask children to build a rectangle with these requirements:
  - ◆ It has only two colors: color A and color B.
  - ◆ It has more than three cubes.
  - ◆ One-third of the cubes are color A and the rest are color B.
- ◆ Have volunteers share their figures and explain how they have satisfied the requirements. Ask them how many cubes of color A they used and total number of cubes they used in the figure. Write the fraction on the board.

On Their Own

How many rectangular Snap Cube shapes can you make that show one-third?

- Work with a group. Use Snap Cubes to create as many rectangles as you can that follow these rules.

  - They have only two colors of cubes.
  - One-third of the cubes are one color.
  - Two-thirds of the cubes are the other color.
  - They have up to 24 cubes.

- Copy each rectangle onto grid paper and color in the squares that show one-third. Then cut the rectangles out.

- Sort the cut-out rectangles into categories according to the total number of cubes you used.

- Make a display of all the rectangles that your group found.

The Bigger Picture

Thinking and Sharing

Have each group post and share their displays. Talk with children about the similarities and differences in the displays.

Use prompts such as these to promote class discussion:

- What patterns do you notice?

- How can you tell that this rectangle shows one-third?

- Look at all the rectangles that use (give a number of cubes, such as 15). How are they alike? How are they different?

- When you say that (give a fraction, such as 4/12) are yellow, what do the 4 and 12 represent in the rectangle? Why is 4/12 another name for 1/3?

- What method did you use for finding all the rectangles that show one-third?

- Did anyone use another method? Describe it.

- Can you think of a number of cubes that makes it impossible to show thirds? Why can't you show thirds with this number of cubes?

Drawing and Writing

Have children explain why they can show one-third with 18 Snap Cubes but not with 16 Snap Cubes. Ask them to use drawings to illustrate their answers.

Extending the Activity

1. Have children repeat the activity for other fractions, such as 1/4, 1/5, or 1/6.

In this discussion, the colors used are blue for 1/3 of the cubes and yellow for 2/3 of the cubes.

Where's the Mathematics?

This activity demonstrates to children that a number, such as 1/3, can have multiple representations. Although the total number of cubes of the rectangle changes from three to six to nine to 12, and so on up to 24, the ratio of blue cubes to the total number of cubes remains one to three. The activity also reinforces children's skills with the multiples of three.

As children find different ways to show one-third, they have the opportunity to learn, demonstrate, and apply some of the basic concepts of fractions: that a fraction is composed of a number of parts that are equal in size, and that the fraction used to name a part of a shape does not depend upon the size of the rectangle. They can appreciate both of those points, for example, by identifying each of the following shapes as showing one-third blue:

      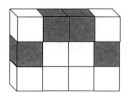

Some children will use proportional reasoning to determine how many cubes they need for a rectangle. These children recognize that for every blue cube, they need two yellow cubes; therefore, for each blue cube they add, they must add two yellow cubes. Other children may gather some cubes and try to separate them into three groups with equal numbers of cubes. For example, six cubes can be separated into three groups of two, whereas seven cubes cannot be separated into three equal groups. Still others will recognize that the both the fractional parts and the total number of cubes needed are multiples of three and choose their cubes by thinking about the multiplication facts.

2. Ask children if it is possible to show thirds with 60 Snap Cubes; with 80 Snap Cubes; with 120 Snap Cubes. If it is possible, ask them to give an example; if not, have them explain why.

When asked to explain why 4/12 is equivalent to 1/3, some children may break the cubes into four groups of one blue and two yellow cubes and explain that one out of every three cubes is blue.

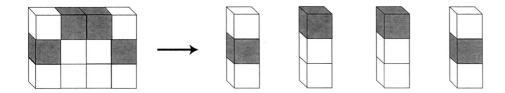

Other children may separate the 12 cubes showing one group of four blue cubes and two groups of four yellow cubes and then explain that one out of every three groups is blue.

It will be beneficial for the entire class to consider the fractional idea of one-third over a period of time (several class periods during which they find as many different ways to display and record one-third as they can). Once children have a chance to study, discuss, and reflect on how various representations show thirds, they will be more ready to move on to other fractions such as 3/5, or 1/6. A deep, conceptual understanding about how fractions are built and recorded will help them in their later work with the abstract forms of computation involving fractions.

TAKE THE CAKE

- Surface area
- Perimeter
- Estimation
- Pattern recognition

Getting Ready

What You'll Need

Snap Cubes, 40–50 in three different colors per pair

Overview

Children use Snap Cubes to build models of one-layer sheet cakes. Then they determine the number of pieces with icing on one, two, or three sides. In this activity, children have the opportunity to

- ◆ link multiplication with the concept of area
- ◆ search for patterns
- ◆ use patterns to make predictions

The Activity

If children have difficulty seeing different kinds of pieces, suggest that they rebuild the cake using one color to represent the pieces that have icing on three sides, a second color to represent the pieces with icing on two sides, and a third color to represent the pieces with icing on one side.

Introducing

- ◆ Show children a 3 x 4 Snap Cube "cake." Explain that each cube represents one serving of cake. Ask them to imagine that the cake is covered with icing and to figure out how many pieces have icing on three sides, how many have icing on two sides, and how many have icing on one side.

- ◆ Once children have reached agreement on the number of each kind of serving, ask a volunteer to post the data on the board.

Size of Cake	Number of Servings	Icing on		
		3 sides	2 sides	1 side
3 x 4	12	4	6	2

On Their Own

How can you find the number of each kind of serving in a square "sheet cake" made of Snap Cubes?

- Work with a partner.

- Use Snap Cubes to build larger and larger square sheet cakes that are all one layer tall. Start with a 2-by-2 cake. Imagine that the top and sides of each cake are covered with icing and that each Snap Cube is one serving.

- For each cake, find the number of servings with icing on three sides, the number of servings with icing on two sides, and the number of servings with icing on one side.

- Keep track of your data in a chart like this:

Size of Cake	Number of Servings	Kinds of Servings		
		Icing on:		
		1 side	2 sides	3 sides
2 x 2	4			

- Look for patterns so that you could describe the number and kinds of servings for any square cake.

The Bigger Picture

Thinking and Sharing

Collect children's data in a class chart. Then discuss the data.

Use prompts like these to promote class discussion:

- ◆ What patterns do you see?

- ◆ How many people could you serve if 25 pieces of the cake have icing only on one side?

- ◆ If you wanted a cake with at least 28 servings with icing on two sides, what size cake would you order?

- ◆ If you made a cake that measured 25 x 25, how many of each kind of serving would there be?

- ◆ How much do you think you should charge for each kind of serving? Using those prices, how much would a 6 x 6 cake cost?

- ◆ If you made a 1 x 1 cake, how much should you charge for that cake? Explain.

Drawing and Writing

Ask children to explain how to find the number of each kind of serving in a 6 x 6 cake. Have them draw a "bird's eye" view of the cake to illustrate their explanation.

Where's the Mathematics?

This activity helps children organize data and look for patterns in a real-world setting. Finding the number of servings in each case reinforces the multiplication facts in a meaningful way.

Size of Cake	Number of Servings	Icing on		
		3 sides	2 sides	1 side
2 x 2	4	4	0	0
3 x 3	9	4	4	1
4 x 4	16	4	8	4
5 x 5	25	4	12	9
6 x 6	36	4	16	16
7 x 7	49	4	20	25

As children explore the square cakes, they may be surprised at first that there are always four servings of cake with icing on three sides no matter how large the cake gets. Some children will recognize that the number of servings with icing on two sides is described by the multiples of four. They may be able to generalize that they can find the number of servings with icing on two sides by multiplying four times two less than the number of pieces on a side. For example, for a 5 x 5 cake, there are 4(5−2), or 12, servings with icing on two sides:

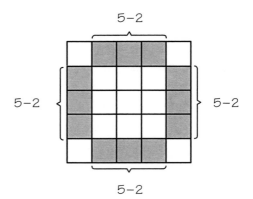

Extending the Activity

1. Have children set a price for each kind of serving and compute the total price for different-sized cakes.

2. Ask children to find the number of each kind of serving in cakes with these dimensions: 4 x 5, 4 x 6, 4 x 7, 4 x 8, and 4 x 9.

The dimensions of a square cake indicate the number of pieces of cake along each edge. All pieces along an edge—except the two corner pieces—have two iced sides. Four times this number gives the total number of pieces with two iced sides. Restating this algebraically, a cake of n x n dimensions would have $4(n-2)$ servings with icing on two sides.

Some children may recognize the pattern of square numbers occurring in the column labeled "Icing on one side." If the children have built the color-coded cakes (as suggested in *Introducing*), it will be easier to see that the square numbers occur because subtracting all the servings around the perimeter of the cake still leaves a square. Some children may be able to generalize that they can find the number of servings with icing on one side by subtracting two from the side and then multiplying the number by itself. Or, a cake of n x n dimensions would have $(n-2)^2$ servings of cake with icing on one side.

Using whatever method they derived from studying the patterns in the data, children can figure out that a 25 x 25 cake will have four servings with icing on three sides, 92 servings with icing on two sides, and 529 servings with icing on the top only. This cake can serve 25^2 or 625 guests.

To guarantee that they get a cake with 28 servings with icing on two sides, children would need to order a 9 x 9 cake. A cake with 25 servings with icing only on the top would come from a cake that measures 7 x 7 and therefore serves 49 people.

For a 1 x 1 cake, there would be a single serving with icing on five sides. Children will have to use some proportional thinking to arrive at a fair price for the cake. Some will multiply the price of "icing only on the top" times five. Others may argue that this cake will only serve one person and charging five times as much is unrealistic. Accept whatever prices the children can justify with a line of reasoning.

The generalizations children make when studying the patterns generated in this activity set the stage for deriving and using formulas.

TETRA FILL-IN

- Spatial visualization
- Game strategies
- Transformational geometry

Getting Ready

What You'll Need

Snap Cubes, 20 of the same color for each team

Snap Cube grid paper, page 90

Tetra Fill-In game boards, page 93

Overhead Snap Cubes (optional)

Snap Cube grid paper transparency (optional)

Overview

Using the five flat shapes composed of four Snap Cubes, children play a game that involves strategically placing the shapes on a grid. In this activity, children have the opportunity to

- ◆ use rotations and flips to fit pieces together
- ◆ develop strategic thinking skills

The Activity

You may want to play part of a game of Tetra Fill-In *with children before they begin the* On Their Own. *If so, prepare two sets of flat tetras, each set a different color.*

Introducing

- ◆ Show children the five flat tetra shapes and a sheet of Snap Cube grid paper.
- ◆ Have a volunteer demonstrate how the tetras can be placed on the grid paper so that the shapes lie within grid lines.
- ◆ Have a second volunteer show how the placement of the tetras can be recorded on grid paper.

On Their Own

Play *Tetra Fill-in!*

Here are the rules.

1. This is a game for two teams of two players each. The object of the game is to fit the last tetra shape on the game board.

2. Each team selects a different color of Snap Cube and builds a set of five tetras.

3.

   The teams take turns placing a tetra on the game board so that the shape fits within the grid lines.

4. Continue playing until no more of the tetras will fit on the game board. You may leave holes on the board that are not big enough for another piece to fit in.

5. The team that places the last tetra is the winner.

6. After the game, record what the board looks like and which team won.

- Play several games of *Tetra Fill-In*.

- Be ready to talk about any winning strategies you find.

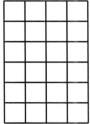

Tetra Fill-In
game board

The Bigger Picture

Thinking and Sharing

After children have had time to play the game several times, call the children together to discuss their observations about the game.

Use prompts such as these to promote class discussion:

- What strategies did your team try?
- What information did you record that helped to find a winning strategy?
- How did you decide which piece to start with?
- How did you decide that a piece would fit?
- How did the other team's choices affect your moves?
- Which shapes did you usually have left at the end of the game?
- Which shapes were easiest to play?
- Did going first make a difference? Why or why not?

Writing

Have children write a paragraph entitled, *How to Win at Tetra Fill-in*.

Extending the Activity

1. Have children determine the least and greatest number of tetras that may be played on a four-by-six game board.

Teacher Talk

Where's the Mathematics?

In discussing the game, children will find it helpful to devise a method for referring to the tetra pieces. For example, they may use the numbers 1–5 to refer to the different shapes, or they may prefer to label them with letters such as T, Z, I, L, and O because the pieces can be rotated or flipped to look like those letters.

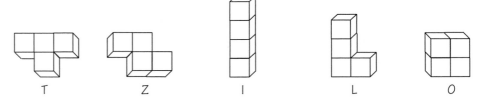

Children will have different ideas about which pieces are hardest to place. Frequently, they identify *Z* and *T* as the hardest ones to place because they have more ins and outs. No matter which pieces children identify as hardest, most children will say that it is best to place these pieces first.

The more children play, the better they become at planning ahead. They can visualize what happens to a piece, rotating and flipping it in their minds before selecting it. They become more skilled in deciding where to place each piece. At first, they may try to place their next piece so it touches a piece previously played. In time, however, they begin to focus on the squares that will be left when a piece is played, realizing that it is a good strategy to place a piece so that the leftover squares do not form one of the tetra shapes.

As they play the game, children will discover that it doesn't matter who goes first; what matters is the piece that is used and where it is placed. For example, in Game 1 below, Team A starts and Team B wins. In Game 2 however, Team A starts and becomes the winner as soon as their square piece is placed in the center at the bottom of the grid. No matter where

---

2. Have children use only one set of tetras to cover the four-by-six game board so that there are no uncovered spaces between the pieces and any uncovered spaces are outside the shape formed by the 20 Snap Cubes. Ask children to determine if there is more than one way to arrange the tetras in this way.

3. Have children play the game with three teams, using three different colors and a larger game board.

Team B places the next piece, it leaves only one more move, which goes to Team A.

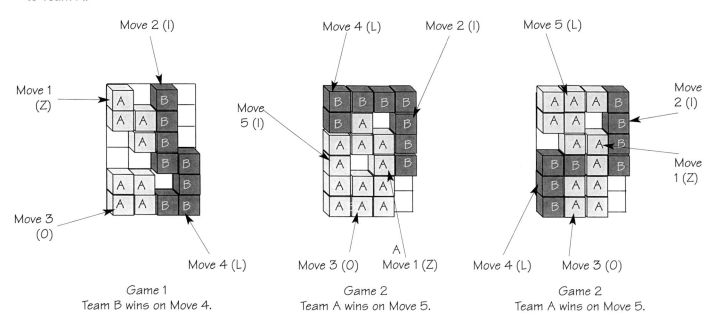

Game 1
Team B wins on Move 4.

Game 2
Team A wins on Move 5.

Game 2
Team A wins on Move 5.

At first, children will play each game to its conclusion. As they become more adept at visualizing what will happen next, they may be able to predict a winner without having to place all the possible pieces.

This activity helps children develop their ability to visualize how shapes fit together, how shapes can be rotated or flipped to fit into specific spots, and what spaces will be left in between shapes as a result of various placements.

THE STAIRCASE PROBLEM

- Organizing data
- Interpreting data
- Multiples
- Looking for patterns
- Computation

Getting Ready

What You'll Need

Snap Cubes, about 65 per pair

Calculators, one per pair

Overview

Children use Snap Cubes to build larger and larger staircases. They predict the number of cubes needed to produce a ten-step staircase that is three cubes wide. In this activity, children have the opportunity to

- ◆ organize and analyze data
- ◆ identify patterns
- ◆ use patterns to make predictions

The Activity

Introducing

- ◆ Display the structures shown and introduce them as "staircases" to the class. Ask children how many steps each staircase has and how wide each step is.

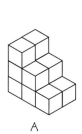
A

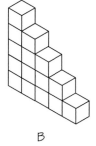
B

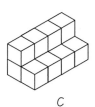
C

- ◆ Ask children to build a copy of staircase A and add another step.
- ◆ After they have added the step to staircase A, ask children to predict how many Snap Cubes they would need to build the next step in staircase A and the total number of cubes in the staircase.
- ◆ Repeat for staircases B and C.

On Their Own

How can you figure out the number of Snap Cubes you need to build a staircase without counting every cube?

- Work with a partner.

- Use your Snap Cubes to build staircases with steps that are three cubes wide. Here are the first three staircases:

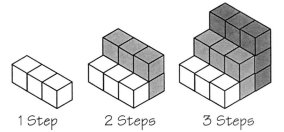

1 Step   2 Steps   3 Steps

- Build the staircases shown above. Then continue to build larger staircases. For each staircase, record how many cubes you use for that step and the number of cubes in the entire staircase.

- Look for a pattern that will help you predict the number of cubes you would need to make a 10-step staircase.

The Bigger Picture

Thinking and Sharing

When all the pairs have finished, call on volunteers to fill in a class chart like this:

Number of Step	Number of Cubes in the Step	Total Number of Cubes in the Staircase
1	3	3
2	6	9
..	..	..
..	..	..

Use prompts like these to promote class discussion:

- ◆ What did you notice as you built your staircase?

- ◆ What patterns did you notice in your data?

- ◆ How did you find the number of cubes you would need to build the 10th step? the total number of cubes in the 10-step staircase?

- ◆ How many cubes would be in the 11th step? How many cubes would be needed to build an 11-step staircase?

Writing

Ask children to write an explanation that would help someone figure out how many cubes would be needed to build a six-step staircase that is four cubes wide.

Extending the Activity

1. Have children explain whether a step in their staircase could have 105 cubes; 500 cubes. If these numbers of cubes could be steps in their staircase, ask children which steps they would be.

Teacher Talk

Where's the Mathematics?

As children complete the problem and discuss their results, they discover that different patterns and problem solving approaches can be used to solve a particular problem. They have the opportunity to extend a pattern to make predictions.

Children may go about building staircases in a variety of ways. Some may build a series of rectangular walls:

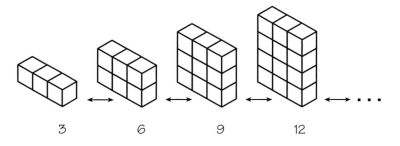

Others may see the stairs as a series of rectangular layers:

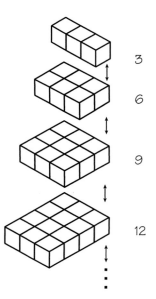

Children may record the numbers of cubes needed to make each step and the total number of cubes for the staircase in a chart like the this one:

2. Ask children to explain whether or not they would double the number of cubes for a 20-step staircase.

3. Ask children to repeat the activity for staircases of different widths.

Number of Step	Number of Cubes in the Step	Total Number of Cubes in the Staircase
1	3	3
2	6	9
3	9	18
4	12	30
5	15	45
6	18	63
7	21	84
8	24	108
9	27	135
10	30	165

Children have only enough Snap Cubes to build a six-step staircase. Therefore, they must look for and use patterns in their data to find information for the 10-step staircase. Children may use different methods for finding data. For example, some children may recognize that the number of cubes in each step is a multiple of three and can be found by multiplying the number of the step by three. Other children may recognize that the entries in the second column can be found by repeatedly adding three.

Similarly, children will have various ways of finding the data in the Total Number of Cubes in the Staircase column. Some children will find the entries in this column by adding the previous entry in the column to the next number in the Number of Cubes in the Step column: 3 + 6 = 9; 9 + 9 = 18; 18 + 12 = 30, and so on. Others might find the entries by adding all of the entries in the Number of Cubes in the Step column: 3 + 6 = 9; 3 + 6 + 9 = 18; 3 + 6 + 9 + 12 = 30, and so on. Still others might find the number of cubes in a staircase only one cube wide (1 + 2 + 3 + 4 + 5 + 6 + 7 + 8 + 9 + 10), and multiply that answer by three.

Once children have figured out the patterns in the table, they should be able to correctly predict that an 11-step staircase will have 33 cubes in the 11th step and 198 cubes in all. They should also be able to answer the first problem in *Extending the Activity:* Since 105 is the 35th multiple of 3, it is possible to have a step with 105 cubes; no step could have 500 cubes since 500 is not a multiple of 3.

The use of calculators allows children to focus on the patterns and concepts involved in the problem, rather than on routine computations.

TRAINS AND BOXCARS

• Multiplication
• Addition

Getting Ready

What You'll Need

Snap Cubes, about 40 per group

Dice, one die per group

Snap Cube grid paper, page 90

Calculators, one per group (optional)

Overview

Children play a game in which they roll a die to determine the number of Snap Cubes to put in a train. They roll the die a second time to determine the number of trains to make. In this activity, children have the opportunity to

◆ view multiplication as repeated addition

◆ practice using multiplication symbolism correctly

◆ add long columns of numbers

4 x 4 = 16

The Activity

You may have to explain what a boxcar is.

You may wish to explain that 4 x 3 = 12 is shorthand for saying four trains of three boxcars equals 12 boxcars.

Introducing

◆ Tell children that you are going to play a game in which Snap Cubes represent boxcars.

◆ Ask for three volunteers. One will be your partner and the other two will be the team you are playing against.

◆ Ask your partner to roll the die. The number that comes up tells how many boxcars are on the train. Have your partner take that number of Snap Cubes and build a train.

◆ Roll the die again to determine how many of those trains to make.

◆ On the chalkboard, record a picture of the trains and the multiplication fact to go with it. For example, if you rolled a 3 and then a 4, the picture would show four trains of three boxcars, and the multiplication fact would be 4 x 3 = 12.

◆ Let the other team play the game and record their trains and multiplication fact on the board.

◆ Compare the two products.

On Their Own

Play *Trains and Boxcars!*

Here are the rules:

1. This is a game for two teams of two players each. In this game, a boxcar is one Snap Cube. The object of the game is to be the team with more boxcars.

2. The first team rolls the die to find out how many boxcars to put on a train. The team builds a Snap Cube train with that number of boxcars. For example, if the team rolls a 3, the team would build a train like this.

3. The first team rolls the die again to find out how many of those trains to build. If that team's second roll is a 4, the team would build a total of four trains with three boxes each.

4. After all the trains are built, the first team draws a picture of the trains and records the multiplication fact that tells how many boxcars there are in all.

5. Then the second team takes their turn.

6. Play continues until each team has had seven turns.

7. Each team finds the sum of all the boxcars from the seven rounds. The team with the greater number wins.

• Save your recording sheets for class discussion.

The Bigger Picture

Thinking and Sharing

Call children together with their recording sheets to discuss the mathematics.

Use prompts like these to promote class discussion:

- ◆ What is the smallest number of boxcars you could get in one turn? How could you get that?
- ◆ What is the greatest number? How could you get that?
- ◆ Did any team get four (10, 16, 20) boxcars in one turn? How did you get that many?
- ◆ Did anyone get any odd numbers of boxcars in one turn? How did that happen?
- ◆ What are the possible numbers of boxcars you can get in one turn? Explain.

Writing

Ask children to write a description of all the ways they could get 12 boxcars in one turn of play.

Extending the Activity

1. Have children play the game using non-standard dice. You might affix stickers to each side of a die with larger numbers written on them. You could also use tetrahedra dice (4-sided) or decahedra dice (10-sided).

Where's the Mathematics?

This activity provides an introduction to the concept of multiplication as repeated addition. Children link pictures of the concrete manipulatives with the symbolic way to represent this repeated addition with a multiplication equation.

Children will find that the smallest number of boxcars is one (1 x 1) and the largest number is 36 (6 x 6), but not every number between 1 and 36 is a possible product. This list shows all of the possible outcomes:

ROLL OF DICE	PRODUCT	ROLL OF DICE	PRODUCT
1, 1	1	4, 1	4
1, 2	2	4, 2	8
1, 3	3	4, 3	12
1, 4	4	4, 4	16
1, 5	5	4, 5	20
1, 6	6	4, 6	24
2, 1	2	5, 1	5
2, 2	4	5, 2	10
2, 3	6	5, 3	15
2, 4	8	5, 4	20
2, 5	10	5, 5	25
2, 6	12	5, 6	30
3, 1	3	6, 1	6
3, 2	6	6, 2	12
3, 3	9	6, 3	18
3, 4	12	6, 4	24
3, 5	15	6, 5	30
3, 6	18	6, 6	36

The possible products, listed in order from smallest to greatest are: 1, 2, 3, 4, 5, 6, 8, 9, 10, 12, 15, 16, 18, 20, 24, 30, and 36.

When asked if every number between 1 and 36 is a possible product, children probably will not think to make the exhaustive list shown above.

2. Pool all of the class' products and make a frequency distribution graph of the products that came up in one game.

Some children may mentally consider a list of products from 1 through 36 and think about whether each of the numbers on that list could be a possibility. After some thought, they are apt to respond, "7 isn't possible" or "17 isn't possible." Other children may suggest making a tally list by asking, "If you got a 7, raise your hand" and continue down the list of possible products from 1 through 36 this way.

Using the data the children collected while playing the game and exploring the different ways some products, such as 4, 10, 16, and 20, could arise touches upon probability and also contributes to building number sense. For example, 4 could arise as two trains of two boxcars or one train of four boxcars or four trains of one boxcar while 16 could only happen one way: four trains of four boxcars. Looking at how the order of the numbers changes from one train of four boxcars to four trains of one boxcar while the product stays the same gives children experience with the Commutative Property of Multiplication.

This game can be a worthwhile activity to leave in a math center so children may return to it later. It also can be suggested to parents, who ask, "What can we do to help our child in math?" When playing the game at home, manipulatives, such as lima beans instead of Snap Cubes, could be suggested.

Children will probably notice that even products occur more often than odd products. Some children may look for patterns to explain why this happens and realize that:

$$\begin{array}{lll} \text{even x even} & = & \text{even product} \\ \text{even x odd} & = & \text{even product} \\ \text{odd x odd} & = & \text{odd product} \\ \text{odd x even} & = & \text{even product} \end{array}$$

Since each die has three even numbers (2, 4, 6) and three odd numbers (1, 3, 5), the chances of getting an even or odd number are the same, and so the odd products will only show up one-quarter of the time. The list of the possible products on the preceding page shows that odd products only show up nine times out of 36, or one-quarter of the time, as predicted. This analysis of even and odd products may help some children when they memorize their multiplication fact tables.

Spending plenty of instructional time on the concept of multiplication will pay off later when children have a mental image of 4 x 3 and the ability to reconstruct this fact as adding four trains of three boxcars.

TWO-COLOR TETRAS

- Congruence
- Spatial visualization
- Symmetry
- Permutations

Getting Ready

What You'll Need

Snap Cubes, 30 each of two different colors per pair

Snap Cube grid paper, page 90

Overview

Children explore all the possible two-color arrangements of the five flat structures that can be made using four Snap Cubes. In this activity, children have the opportunity to

- ◆ develop spatial sense by flipping and rotating shapes
- ◆ explore different kinds of symmetry
- ◆ explore strategies for identifying all of the possible arrangements

The Activity

If children need practice in comparing arrangements, show the arrangement on the left and rotate and flip it until it matches the arrangement on the right. Explain that since this can be done, the arrangements are not different.

Introducing

- ◆ Show these two-color arrangements made of three Snap Cubes. Ask children to try to build other L-shaped arrangements in two colors.

- ◆ Have volunteers offer solutions.

- ◆ Establish that there are only four different arrangements. Rotate and flip these arrangements to confirm that they are different.

- ◆ Discuss ways to record the four different arrangements on grid paper. You may wish to color the arrangements or just record the initial letter of the color.

On Their Own

> *How many different tetras can you make using only 2 colors of Snap Cubes?*
>
> - Work with a partner to build a tetra. A *tetra* is a structure made of four Snap Cubes. Follow these rules:
>
>   - Your tetra must use two colors.
>
>   - Your tetra must be flat. That is, you could place the structure on a table so that all four Snap Cubes would touch the table.
>
>
>
>   Okay   Not okay
>
> - Each of you should build the same tetra shape again but try to change the arrangement of the two colors.
>
> - Compare your tetras. If they have the same color arrangement, keep only one. If they have different color arrangements, then keep both of the tetras.
>
> - Continue building until you can't make any more color arrangements that are different from what you already have.
>
> - Record your findings.
>
> - Repeat the activity with another tetra shape.
>
> - Be ready to discuss how you know you have found all the 2-color arrangements.

The Bigger Picture

Thinking and Sharing

Invite a pair of children to choose one tetra, display all the two-color arrangements they have found, and discuss their methods and results. Ask the class to verify that no other arrangements of that tetra exist. Then repeat the procedure for other tetras.

Use prompts like these to promote class discussion:

- How did you compare tetras?

- What strategies did you use to decide whether you had found all of the possible arrangements?

- How can you convince someone that you have found all of the possible arrangements?

- Which shapes have the most possible arrangements? Which have the fewest?

Writing

Ask children to select the shape with the fewest possible two-color arrangements. Ask them to explain why the shape has so few arrangements.

Teacher Talk

Where's the Mathematics?

Pairs of children may find some or all of the 44 possible tetras with different two-color arrangements.

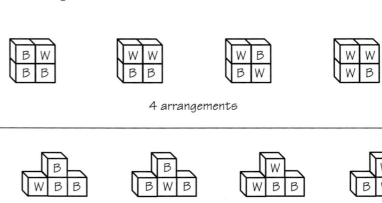

4 arrangements

10 arrangements

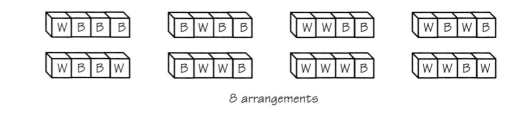

8 arrangements

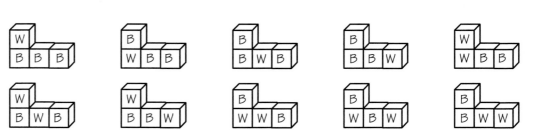

14 arrangements

Extending the Activity

Have children use the shapes for the other three tetras (see cards F, G, and H on page 92) and make all of the possible arrangements with two colors.

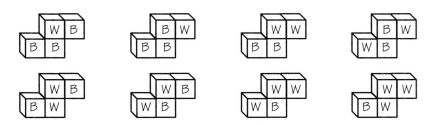

8 arrangements

As they search for all of the possible arrangements of a given shape, children have the opportunity to develop their ability to visualize shapes in different positions. Through this process, children can gain an intuitive understanding that shapes that have been flipped or rotated and still look alike are congruent—although they are in a different position, they will still have the same shape.

Many children will approach the task of finding all possible two-color combinations in a very random manner. Other children may develop a system for finding all the possible configurations. For example, they may start with three black cubes and one white cube. They may then move the one white cube to find all the possible positions in which it could be placed. Then they may explore all the combinations possible with two white cubes and, finally, with three white cubes. Some children may use the idea of reversing colors. That is, wherever they had a white cube, they will make it black and wherever they had a black cube, they will make it white to find additional combinations.

Some children may use the idea of symmetry to explain why different shapes have different numbers of arrangements. The square tetra has the fewest possible arrangements because it has the most lines of symmetry. As the number of lines of symmetry decreases, the number of arrangements increases. The L-shaped tetra which does not have any symmetry has the greatest number of arrangements (14).

Some children may also recognize that the numbers of arrangements are all even numbers. If they have found seven arrangements, this idea of even numbers may give them the persistence to search for that last elusive arrangement. They may also notice that, for each shape, combinations of two white cubes and two black cubes always gives the greatest number of arrangements. Combinations using one white cube have the same number of arrangements as combinations using three white cubes.

The ability to find and organize combinations is an important skill that children will use in the study of probability.

WRAP IT UP

- Surface area
- Volume
- Spatial visualization
- Pattern recognition

Getting Ready

What You'll Need

Snap Cubes, 24 per pair
Calculators, one per pair (optional)

Overview

Children use Snap Cubes to build different rectangular prisms and find the surface area of each. In this activity, children have the opportunity to

- ◆ work with three-dimensional figures
- ◆ discover that shapes with the same volume can have different surface areas
- ◆ search for patterns

The Activity

If children have difficulty keeping track of the number of Snap Cube faces, suggest that they count the number of holes and posts that show. Another method would be to place sticky dots on the faces as they count them.

You may wish to review what is meant by the dimensions of a rectangular prism.

Introducing

- ◆ Have children examine a Snap Cube. Establish that it has six square faces.
- ◆ Ask children to predict how many square faces will be visible when they snap two cubes together.
- ◆ Tell children that counting the visible Snap Cube faces tells how much surface area a figure has. Explain that surface area is the number of square units that are on the outside of a figure and that it can be useful to people who design the packaging for various products. For example, if one Snap Cube represented a piece of candy, six square units of paper would be needed to wrap it up. Likewise, 10 square units of paper would be necessary to wrap two pieces of candy in one package.

On Their Own

> **Can you help a candy company figure out how much paper they would need to cover boxes holding certain numbers of candies?**
>
> - Work with a partner. Imagine that a Snap Cube represents a box that holds one piece of candy.
>
> - Using Snap Cubes, build all the rectangular boxes that can hold eight candies. Your boxes may include more than one layer of candy.
>
> - For each box, count the number of Snap Cube faces on the sides to find out how many square units of paper would be needed to cover the box.
>
> - Record the number of pieces of candy, the dimensions of the boxes, and the number of square units of paper needed to cover each box.
>
> - Repeat the activity for boxes holding 12 candies and then for boxes holding 24 candies.
>
> - Look for patterns in the data you recorded.

The Bigger Picture

Thinking and Sharing

Collect the data in a class chart with headings *Number of Pieces, Dimensions of Box,* and *Amount of Paper.* Discuss the data.

Use prompts like these to promote class discussion:

- What patterns do you see?

- What methods did you use to find the amount of paper needed?

- Which packages require more paper: the long, skinny packages or the more compact ones? Explain.

- Why might a candy company want to use as little paper as possible?

- Why might a candy company want to use as much paper as possible?

If children are ready for the vocabulary, you may want to rename the headings Volume, Dimensions, *and* Surface Area *in the class chart.*

Writing

Have children write a note to the candy company president describing how they think the company should package 24 candies and why. Ask them to include a description of the dimensions of the package, type of material to be used, and ideas for the wrapper.

Extending the Activity

1. Ask children to find how many candies could be packaged in 30 square units of paper.

2. Have children investigate the surface area of an 8-cube structure if the cubes do not have to be arranged in rectangular prisms.

Where's the Mathematics?

This activity helps children see that packages, or rectangular prisms, with the same volume can have different surface areas. The more compact the package is, the less surface area it has. If a company wants to save money on packaging, it would choose the compact package. On the other hand, a candy company might favor a long, skinny package because it makes the consumer think that there is more candy in the package because it is "spread out."

As children explore the ways to package the candy, some of them will attack the problem using trial and error, taking the specified number of cubes and randomly rearranging them until they form a rectangular prism. Others will have a more orderly approach, first searching for all the prisms one cube high, then two cubes high, and so on.

VOLUME	DIMENSIONS	SURFACE AREA
8	1 x 1 x 8	34
8	1 x 2 x 4	28
8	2 x 2 x 2	24
12	1 x 1 x 12	50
12	1 x 2 x 6	40
12	1 x 3 x 4	38
12	2 x 2 x 3	32
24	1 x 1 x 24	98
24	1 x 2 x 12	76
24	1 x 3 x 8	70
24	1 x 4 x 6	68
24	2 x 2 x 6	56
24	2 x 3 x 4	52

When asked to identify patterns, some children may notice that all of the surface areas are even numbers. When asked to explain why that is so, they might show how the top and the bottom have the same surface area, the back and the front have the same surface area, and the two sides also have matching surface areas; so each number was doubled. Some children might point out that organizing their dimensions in an orderly fashion helped them know that they had found all the possible ways to package the candy. Some children may express surprise that doubling one of the dimensions did not necessarily double the surface area. For example, a 1 x 1 x 12 package has a surface area of 50. Doubling the first dimension and making the package 1 x 2 x 12 only has 76 square units of surface area, which is considerably less than 2 x 50.

When asked to explain their methods of finding the surface area, some children may report that they counted each hole and post. Others will say that they found the surface area of the top and used the same number for the bottom, and then did the same for the front and back and the two sides. They may report adding up the surface areas of the six faces or they may talk about multiplying by two. Some children will describe that when they found the surface area of the 2 x 2 x 3 prism, for example, they saw 8 rows of 3 holes as they looked around the top, bottom, and two sides and four holes on each of the two ends.

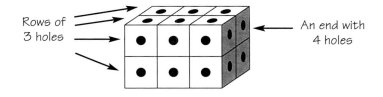

Rows of
3 holes

An end with
4 holes

CLEARED FOR TAKE-OFF
GAME BOARD

				1	1					
				2	2					
				3	3					
				4	4					
				5	5					
				6	6					
				7	7					
				8	8					
				9	9					
				10	10					
				11	11					
				12	12					

TEAM A

TEAM B

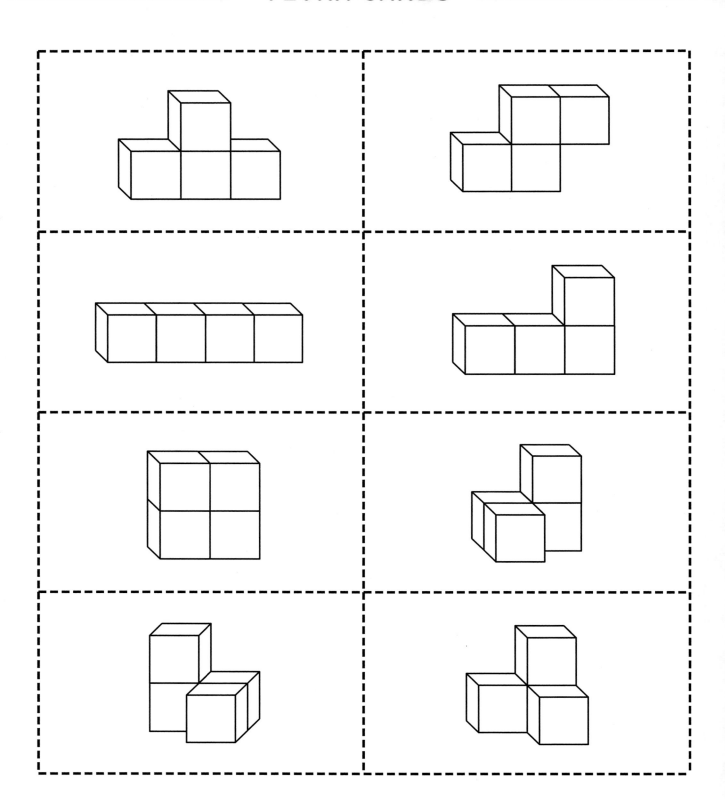

ISOMETRIC DOT PAPER